Zero-Symmetric Graphs

Trivalent Graphical
Regular Representations
of Groups

Academic Press Rapid Manuscript Reproduction

Zero-Symmetric Graphs
Trivalent Graphical Regular Representations of Groups

H. S. M. COXETER

Department of Mathematics
University of Toronto
Toronto, Ontario, Canada

ROBERTO FRUCHT

Facultad de Ciensias
Universidad Técnica Federico Santa María
Valparaiso, Chile

DAVID L. POWERS

Department of Mathematics and Computer Science
Clarkson College of Technology
Potsdam, New York

ACADEMIC PRESS   1981

A Subsidiary of Harcourt Brace Jovanovich, Publishers
New York   London   Toronto   Sydney   San Franciso

ACADEMIC PRESS, INC.
111 Fifth Avenue, New York, New York 10003

United Kingdom Edition published by
ACADEMIC PRESS, INC. (LONDON) LTD.
24/28 Oval Road, London NW1 7DX

Library of Congress Cataloging in Publication Data

Coxeter, H. S. M. (Harold Scott Macdonald), Date
    Zero-symmetric graphs.

    Includes index.
    1. Graph theory. 2. Representations of groups.
I. Frucht, Roberto. II. Powers, David L. III. Title.
QA166.C66   511'.5   81-4604
ISBN 0-12-194580-4

Dedicated to Ronald M. Foster

Contents

PART III   GRAPHS OF TYPE ^3Z

Preface

At the Conference on Graph Theory and Combinatorial Analysis held at the University of Waterloo in 1966, Ronald Foster presented a *Census of trivalent symmetrical graphs,* a draft of which was distributed to a dozen colleagues. (''Symmetrical'' means edge-transitive as well as vertex-transitive.) In the same year, in a letter to the first author of this book, Foster suggested the study of those finite trivalent graphs whose automorphism group acts regularly on the vertices, coining for them the term ''0-symmetric.'' Loosely speaking, these are trivalent graphs that are just vertex-transitive, in the sense that they have no further symmetry.

In 1975, in *Notes* distributed again only to a reduced number of friends, he began the study of these 0-symmetric graphs and also of the ''*t*-symmetric'' graphs, which represent an intermediate class between the 0-symmetric and the symmetrical. In particular, he studied the most numerous family of 0-symmetric graphs, those whose automorphism group is isomorphic to a dihedral group. In Table 22.1 we list, from Foster's work, the 350 graphs of this type having not more than 120 vertices (the upper limit we have fixed, somewhat arbitrarily, for this study). For these and other contributions, the authors dedicate this book to him.

We also wish to acknowledge the contributions of Mark Watkins, who found the first examples of the graphs studied in Sections 23 and 24.

From the preceding it should already be clear to the reader that the aim of this book is to describe all of the 0-symmetric graphs with not more than 120 vertices that we have found during several years of intensive search. In spite of our intentions, we very likely have overlooked some 0-symmetric graphs or erroneously included some that are not 0-symmetric because of a hidden symmetry. We will be most grateful if a reader finding any omission or error would communicate the facts to any or all of the authors.

The work of the first author was supported in part by Canada's Natural Sciences and Engineering Research Council, Grant No. A2338. The work of the third author was supported by the Organization of American States, through Fellowship BEGES 62564, during a sabbatical leave at Universidad Santa María.

H.S.M.C.   *University of Toronto*
R.F.   *Universidad Técnica F. Santa María.*
D.L.P.   *Clarkson College of Technology*

Part I Generalities

1   INTRODUCTION

The graphs considered in this book are finite, connected, vertex-transitive and trivalent. In a rather natural way, these graphs can be divided into three classes, according to the number of "essentially different" edges incident at each vertex. Two adjacent edges are essentially different if there is no graph automorphism taking one into the other while fixing their common endpoint. If such an automorphism does exist, the two edges are considered to be "alike". The three classes, which we call S, T and Z, are described below.

Class S. All three edges incident at any vertex are alike; that is, the graph is both edge- and vertex-transitive. Foster (1966) made a census of these symmetrical graphs with up to 400 vertices. Symmetrical graphs can be further subdivided. The order of the stabilizer of any vertex (that is, the ratio of the order of the automorphism group to the number of vertices) must be of the form $3 \cdot 2^{s-1}$, with s not greater than 5 (Tutte, 1947). This fact gives rise to the name s-regular (or s-unitransitive* (Harary, 1969)). For example, since we do not admit "multigraphs", the smallest graph in class S is the complete graph on four points, K_4. Since its automorphism group is the symmetric group S_4, of order 24, the graph is 2-regular.

---

*Other deviations from the terminology of Harary (1969) are the terms vertices, edges, valency, trivalent and hamiltonian circuit instead of points, lines, degree, regular of degree 3, and hamiltonian cycle.

<u>Class T</u>. There are two kinds of essentially different edges at each vertex. Thus if vertex v_0 is adjacent to v_1, v_2 and v_3, then there exists a graph automorphism leaving v_0 fixed and interchanging, let us say, v_1 and v_2, but none taking v_3 into v_1 or v_2. It is easy to see that the stabilizer of any vertex has order 2^t ($t \geq 1$), and the graphs of this class will be called t-<u>symmetric</u>.

The smallest graph in class T is the triangular prism. (When the name or symbol for a polyhedron is used to designate a graph, we mean the "skeleton": just the vertices and edges, ignoring the faces—though usually the faces are significant as short circuits belonging to the graph.) At each vertex there are two edges, each lying on one square and one triangle, and one edge that lies on two squares. The automorphism group of the graph is the dihedral group D_6 of order 12, and the stabilizer of any vertex is the cyclic group C_2 of order 2. Thus the graph is 1-symmetric.

<u>Class Z</u>. There are three kinds of essentially different edges incident at each vertex. In this case, the stabilizer of any vertex is trivial—that is, contains only the identity. Since this corresponds formally to the limiting case $t = 0$ for the order of the stabilizer of a t-symmetric graph, we shall follow Foster (see the Preface) in calling these graphs 0-<u>symmetric</u>. Evidently, for each pair of vertices in the graph there is a unique automorphism which carries the first into the second.

The smallest graph in class Z, shown in Fig. 1.1, has 18 vertices. The 0-symmetry of this graph is by no means

obvious, although it can be proved by means of theorems given
in Section 4.

   This book will be devoted to the study of graphs of class
Z.   We describe and use a variety of methods for obtaining
0-symmetric graphs, in an attempt to compile a census of
those having not more than 120 vertices.

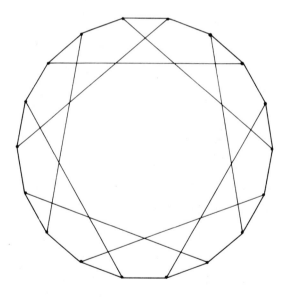

Fig. 1.1   The smallest zero-symmetric graph.

2   CAYLEY GRAPHS (IN GENERAL)

It is well known that Cayley (1878a, 1878b) gave the following visualization of the multiplication table of a finite group H by what is called today a Cayley diagram. (He called it a color-group.) This is a digraph having one vertex for each element of H, where it is convenient to use the same symbol for an element and its corresponding vertex. With each generator R_i of H we associate a certain set of directed and colored edges, the direction indicated by an arrow and the subscript i of the generator by using the ith color. Two vertices, P and Q, are joined by an edge of the ith color and directed from P to Q, whenever

$$PR_i = Q . \tag{2.1}$$

It is easy to see that if n_1 is the period of R_1, the edges showing the first color will form cyclically directed n_1-gons (and similarly for the other colors). However, to avoid 2-circuits consisting of two oppositely directed edges joining the same two vertices, we shall agree once and for all to use a single undirected edge whenever a generator is involutory (i.e., of period two).

It should be pointed out that redundant generators are allowed; it is only forbidden to include the identity E of H in the set of generators or to use a non-involutory element together with its inverse as generators. So, for instance, the four-group $D_2 \simeq C_2 \times C_2$ (in terms of two generators A and B) admits as Cayley diagram a square EACB in which the edges EA, AC, CB, BE are alternately colored red and blue (say); but the diagonals EC and AB, with a third color, might be

6

added, since C = AB might be used as a redundant generator.
Further instances of Cayley diagrams are to be found in
Coxeter and Moser (1980), Fig. 3.3 or Grossman and Magnus
(1964), Fig. 12.1; see also Figs. 3.1 and 3.2 below.

It is easy to see that the permutations of the vertices
which preserve the colors of incident edges are those given
by the so-called left regular representation of the group H.
More precisely, when the group acts on the diagram, its
element P takes the vertex P to E, not vice versa. However,
once the Cayley diagram has been drawn, the names of the
vertices have no more importance and can be dropped, and we
still have a faithful representation of the given group H.

Such a diagram is, of course, not an ordinary graph in
the sense in which that term is used by Harary (1969), since
the edges are colored and (if there are non-involutory generators) directed. Hence in order to transform a Cayley
diagram of a group H into an ordinary graph G, we may drop
the direction arrows and omit colors. In accordance with
general usage the graph thus arising from a Cayley diagram of
a group H will be called a Cayley graph of H with respect to
the generators used. (So both the square {4} and the complete graph K_4 are Cayley graphs for D_2, as we have seen
above).

Since such Cayley graphs are obviously always vertex-
transitive (a proof is given by Biggs (1974), Proposition
16.2.1) and sometimes also edge-transitive, this procedure
has already been used—for instance by Frucht (1955) and more
recently Coxeter (1973)—to obtain interesting vertex-
transitive graphs. It will be used here as the main tool for
obtaining 0-symmetric graphs.

It should however be pointed out that in general the automorphism group $\Gamma(G)$ of a Cayley graph G for a group H is strictly larger than H itself, so that G cannot be considered as a faithful representation of H. For instance, in the case considered above of the four-group D_2 we had {4} and K_4 as Cayley graphs, but $\Gamma(\{4\})$ is the dihedral group D_4 or order 8, and $\Gamma(K_4) \cong S_4$ of order 4! = 24. For another example see Biggs (1974), p. 107.

So the groups H having a Cayley graph G such that $\Gamma(G) \cong H$ seem to be of special interest and have already been considered by Watkins and others (Imrich 1975; Nowitz and Watkins 1972a, 1972b; Watkins 1971, 1974a) who have called them groups "admitting a GRR" (= graphical regular representation). For instance, the dihedral groups of order at least 12 have such a "GRR" (but not those of lower order). In general, however, the corresponding graphs are not trivalent—the only case in which we are interested in the rest of this book.

3   TRIVALENT CAYLEY GRAPHS; THE LCF AND FRUCHT NOTATIONS

From now on, only trivalent and connected vertex-transitive graphs will be considered—as we already did in the Introduction where the three classes S, T, and Z were defined.

It will be convenient to introduce a subdivision of these classes in subclasses or types, according to the possibility of obtaining the graphs under discussion as Cayley graphs (in the sense discussed in Section 2).

If it is impossible to obtain a given trivalent vertex-transitive graph as a Cayley graph, the prefix N will be used. For example, the Petersen graph belongs to NS (or "is NS" as we'll say for short; a proof of this fact is sketched by Biggs (1974), pp. 107-108). If a trivalent graph is the Cayley graph of some group H, there are obviously only the following two possibilities as to the number of generators used in the construction of the Cayley diagram: either we use <u>one</u> involutory generator and one of period $\geq$ 3, or we use <u>three</u> involutory generators. We distinguish these two cases by using the presuperscript 1 or 3, respectively (corresponding to the number of involutory generators). For instance, K_4 is ^3S, being the Cayley graph of the four-group $D_2 \cong C_2 \times C_2$ when generated by its three involutory elements (a possibility that was already mentioned above). However, the same graph K_4 is also ^1S as the Cayley graph of the cyclic group C_4 if we use one element of period 4 together with its square as generators in the Cayley diagram. (The involutory generator is of course redundant as a generator of the group C_4, but we need it to have a <u>trivalent</u> Cayley graph.)

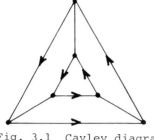

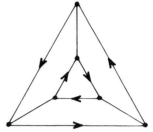

Fig. 3.1 Cayley diagram   Fig. 3.2 Cayley diagram
          of C_6.   of $D_3 \cong S_3$.

As the smallest instance of a graph that is ^1T we have T_3,
the triangular prism. Indeed, T_3 is a Cayley graph of both
the groups C_6 and $D_3 \cong S_3$, when we generate them by one
element of period 3 and one involutory element; see Figs. 3.1
and 3.2 for the corresponding Cayley diagrams.

It is not known to the authors whether Cayley graphs or
at least 0-symmetric graphs are always hamiltonian. However,
it turns out to be true for almost all the graphs in this
survey. A convenient device for the concise description of
a trivalent hamiltonian graph is the so-called LCF notation
explained by Frucht (1977).

Label the p vertices of the graph in such a way that
v_0v_1, $v_1v_2, \ldots, v_{p-2}v_{p-1}$, $v_{p-1}v_0$ are the edges of a
hamiltonian circuit; then (thinking of the hamiltonian circuit as a closed p-gon) each vertex v_i is joined by a
"diagonal" or a "chord" to just one other vertex v_j that
comes on the polygon d_i steps later: that is,

$$j \equiv i + d_i \quad (\text{mod } p), \quad 2 \leq d_i \leq p-2 . \qquad (3.1)$$

The collection $[d_0, d_1, d_2, \ldots, d_{p-1}]$ of the p numbers d_i
gives a practical (although not unique) description of a

trivalent hamiltonian graph, which can still be improved by the
following two rules:

Rule 1.   If any $d_i > p/2$, replace it by $d_i - p$.

Rule 2.   Avoid repetitions of strings of d_i in the case of
periodicity, using a self-explanatory exponential notation.

For instance, the cube can be labeled as shown in Fig. 3.3.
Then we see that the d_i are

$$
d_i = \begin{cases} 3, \text{ for } i = 0, 2, 4, 6 \\ 5, \text{ for } i = 1, 3, 5, 7 \end{cases}
\tag{3.2}
$$

By applying Rule 1 we obtain the code $[3,-3,3,-3,3,-3,3,-3]$,
which reduces to the shorter form $[3,-3]^4$ upon application of
Rule 2.

It is hardly necessary to point out that, since the starting point on the hamiltonian circuit is immaterial, the numbers
d_i in the LCF code might be permuted cyclically; so the cube
might be equally well described by the code $[-3,3]^4$.   Sometimes
this possibility—and also that of choosing between essentially
different hamiltonian circuits— will be used below to obtain
the "nicest" LCF notation for a 0-symmetric graph.   In

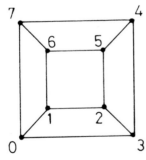
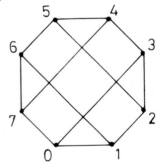

Fig. 3.3   Two drawings of the cube labeled to show
          a hamiltonian circuit.

particular, preference will be given to "antipalindromic" codes, those in which

$$d_{p-1-i} \equiv -d_i \pmod{p}, \quad i = 0,1,\ldots, \frac{p}{2} - 1 \,. \tag{3.3}$$

Indeed, so many of the codes are antipalindromic that we shall note only the absence of this property. When it is convenient, we abbreviate an antipalindromic code by replacing the second half with the symbol ";-". For instance the LCF code

$$[18,9,-17,29,-9,18,-18,9,-29,17,-9,-18]^5 \tag{3.4}$$

will be abbreviated by

$$[18,9,-17,29,-9,18;-]^5 \,. \tag{3.5}$$

Since the absolute values of the d_i (when applying Rule 1) might be interpreted as chord lengths in the hamiltonian circuit, one might be tempted to use the number N_c of different chord lengths appearing in the "nicest" LCF code as a method for classifying vertex-symmetric graphs in general and 0-symmetric ones in particular. For instance, the cube, where all the $|d_i|$ have the same value 3, has $N_c = 1$, and the same holds for the graph K_4 described by the code $[2]^4$, while higher values of N_c might be thought of corresponding to a higher "complexity" of the graph.

REMARKS: (i) A hamiltonian circuit in a Cayley graph obviously corresponds to a relation

$$R_\alpha R_\beta R_\gamma \ldots R_\omega = E \tag{3.6}$$

in the generators R_i, where the number of factors on the left-hand side equals the number p of vertices of the graph, and where the p products

$$R_\alpha, \ R_\alpha R_\beta, \ R_\alpha R_\beta R_\gamma, \dots, \ R_\alpha R_\beta R_\gamma \dots R_\omega \tag{3.7}$$

are all different (thus representing all the elements of the
group). Any relation of this kind will lead to a hamiltonian
circuit and hence to an LCF notation for the Cayley graph.
For this reason we shall give not only the LCF codes, but also
the relations from which they were derived.

(ii) From an LCF code it can be immediately seen whether
a trivalent hamiltonian graph is bipartite or not; indeed it
is bipartite if and only if all the d_i are odd.

For the girth, however, we have only the inequality

$$g \leq 1 + \min(|d_o|, |d_1|, |d_2|, \dots, |d_{p-1}|) \ . \tag{3.8}$$

Since most of the graphs in this study were found to have
a hamiltonian circuit, we have described them by means of LCF
codes whenever feasible. In the few cases in which we were
unable to do this, we have provided some other method by
which the graph can be constructed—for instance, giving a
permutation representation for the generators of the group.

In many cases we have found it helpful to use a device
that allows both visual representation and parametrization of
a graph or a family of graphs. It was proposed by Frucht
(1970) and has been described by Capobianco and Molluzzo
(1978, Appendix) under the name of "Frucht notation". We
explain it here only for the special case of a trivalent
Cayley graph G.

Let ϕ represent the graph automorphism of left multiplication by some element $X \neq E$ of H. Then the orbits of ϕ, say

O_1, O_2,...,O_k, each contain the same number of elements,
since they represent the right cosets of the cyclic group (of
order m) generated by X.

From each orbit we choose a representative, say v_α for
O_α, and then refer to the elements of O_α as $\phi^i(v_\alpha)$, i = 0,1,
...,m-1. In the Frucht notation, each orbit is represented
by a circle in which we write m, the number of elements it
contains.

If a vertex v_α is adjacent to vertex $\phi^k(v_\beta)$, then also
$\phi^i(v_\alpha)$ is adjacent to $\phi^{k+i}(v_\beta)$ for any i. So this adjacency
can be represented by drawing an arrow from the circle repre-
senting O_α to that representing O_β, with the number k, called
the "lift" of the arrow, in the middle, if k ≠ 0. If k = 0,
both the 0 and the arrowhead may be deleted, the circles corresponding to O_α and O_β being joined by a line. For an
example, Fig. 3.4 shows the Frucht notation for the smallest
known 0-symmetric graph, having LCF code $[5,-5]^9$ and shown in
Fig. 1.1.

It is hardly necessary to point out that an arrow with
lift k may be replaced by an arrow in the opposite direction
with lift m-k. This will be convenient whenever m/2 < k < m.

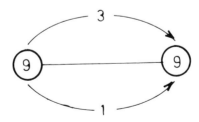

Fig. 3.4   Frucht notation for the graph $[5,-5]^9$. See
           also Fig. 22.2.

Fig. 3.5 Frucht notation for the triangular prism.

In case $\alpha = \beta$ — that is, if v_α is adjacent to $\phi^k(v_\alpha)$ — the number m in the circle will be replaced by m(k), and the loop that would result from following the directions above is suppressed. For example, Fig. 3.5 shows the Frucht notation for the triangular prism.

Several arbitrary choices are involved in the construction above, and changing them also changes the Frucht notation for a graph. First, instead of v_α in circle α, we might choose any other vertex, say $\phi^r(v_\alpha)$, as representative. The effect would be to increase by r the lifts of all arrows going out of circle α and to decrease by r the lifts of those coming in.

Second, instead of automorphism ϕ we might as well choose a power, ϕ^s, if (s,m) = 1. The result is to "reorder" the vertices in each orbit, and requires that each lift k be replaced by k' where

$$k' \equiv s'k \pmod{m} , \qquad\qquad (3.9)$$

and s' is the inverse of s (mod m) — that is

$$ss' \equiv 1 \pmod{m} . \qquad\qquad (3.10)$$

The proof is left to the reader. An example will be found below.

In many cases a hamiltonian circuit in a graph can be found by examining its Frucht notation. Suppose there is a

succession of circles and arrows (or lines without arrows)
that begins and ends at the same circle and passes through
each circle just once. If the algebraic sum of the lifts
encountered is coprime to m, then the corresponding closed
path in the graph is a hamiltonian circuit.

As an example consider the Frucht notation for a graph
with 60 vertices in Fig. 3.6. The clockwise closed path with
lifts -1,0,3,0, whose sum, 2, is coprime to 15, provides a
hamiltonian circuit. To calculate its LCF code it is convenient to pass to another Frucht notation where the algebraic
sum of the lifts equals 1. To this end, use (3.9) above with
s = 2; that is, multiply all lifts by 8 (mod 15). In this way
we obtain Fig. 3.7 or—if we change representatives in the
left-hand side of the diagram—Fig. 3.8. From the latter, the
LCF code $[27,9,-9,-27]^{15}$, corresponding to the graph No. 1 of
Table 20.1, can easily be read off.

Finally, we note that the Frucht notation for the Cayley
diagram of a group bears somewhat the same relation to
Schreier's coset diagram—see Coxeter and Moser (1980), p. 31—
that the Cayley graph bears to the Cayley diagram of the same
group.

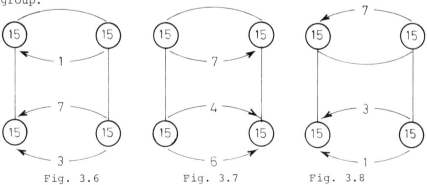

Fig. 3.6   Fig. 3.7   Fig. 3.8

Frucht notation for the Cayley graph of group B(15,4).
(See Table 20.1).

4   GENERAL REMARKS ON 0-SYMMETRIC GRAPHS

There are three ways in which 0-symmetric graphs might be
considered simpler than those of the classes S and T:

(i) 0-symmetric graphs are always Cayley graphs—in other
words, the subclass NZ is empty:

$$NZ = \emptyset \; ; \tag{4.1}$$

(ii) any 0-symmetric graph is the Cayley graph of just one
abstract group, namely of its own automorphism group;

(iii) no 0-symmetric graph can be ^1Z and ^3Z at the same
time:

$$^1Z \cap {}^3Z = \emptyset \; . \tag{4.2}$$

All these facts are immediate consequences of the definition of class Z. They are not only of theoretical interest,
but also of practical importance for the following reasons.

It follows from (iii) that the cases ^1Z and ^3Z can be
dealt with separately, and this will be done in Parts II and
III respectively. From (i) and (ii) it follows that for a
census of 0-symmetric graphs it would be sufficient to make a
systematic search for groups allowing a "graphical regular
representation via trivalent graphs" (in Watkins' terminology),
that is, of groups whose Cayley graphs belong to class Z.

However, this remark should not be misinterpreted to mean
that the only difference between this study and that of
Watkins and others (mentioned above) is our limitation to
trivalent graphs. Indeed, we are interested not only in the
groups giving rise to 0-symmetric graphs (when using suitable

17

generators), but also in the graphs themselves; and it will be
seen below that—depending on the choice of the generators—some
groups can have several non-isomorphic, and sometimes quite
different, 0-symmetric Cayley graphs.

Another consequence of (i) and (ii) is the following.

THEOREM 4.1: There are no 0-symmetric trivalent graphs with
abelian automorphism group.

PROOF: From Imrich (1975) it is known that the only abelian
groups admitting a "graphical regular representation" are C_2
and the direct products $C_2 \times C_2 \times ... \times C_2$ of order at least 32.
None of these groups can however be generated by just three
involutory generators. (It is hardly necessary to point out
that no element of period $\geq$ 3 exists in these groups). #

As to non-abelian groups, of course those already known as
not admitting any graphical regular representation at all
(such as the dicyclic groups) can be disregarded in our study.

Necessary conditions for a group to have a 0-symmetric
Cayley graph will be given below (separately for the cases ^1z
and ^3z). Unfortunately, we have been able to find sufficient
conditions only for special families of groups, such as the
dihedral and some related groups. In other cases we had to
check individually each graph (or at least each family of
graphs) in order to be sure that the graph under consideration
belongs really to class Z, and not to class S or T because of
some "hidden symmetry". Lack of space does not allow us to
give in this book detailed information about those checks.
Let us point out at least that the following theorem has been
most useful in proving 0-symmetry. The statement given is a

modified form of Proposition 2.1 of Nowitz and Watkins (1972a)
and Corollary 1.4 of Imrich (1973; see also Godsil 1979).

THEOREM 4.2:   In the Cayley graph of a finite group H, let K
be any set of vertices, each of which is fixed by every graph
automorphism that fixes the identity E.   If H is generated by
the elements corresponding to K, then the graph is 0-symmetric.
PROOF: Let ϕ be a graph automorphism that fixes the vertex
labeled E, and let P be the label of an element of K.   By λ
we denote the graph automorphism that takes the vertex labeled
X to that labeled PX.   Now

$$(\lambda^{-1}\phi\lambda)E = \lambda^{-1}\phi(PE) = \lambda^{-1}P = E , \qquad (4.3)$$

so $\lambda^{-1}\phi\lambda$ also fixes E.

Let Q be an element of K.   Then the vertex labeled Q is
also fixed by the graph automorphism $\lambda^{-1}\phi\lambda$.   That is,

$$Q = (\lambda^{-1}\phi\lambda)Q = \lambda^{-1}\phi(PQ) , \qquad (4.4)$$

or in other terms

$$PQ = \lambda Q = \phi(PQ) . \qquad (4.5)$$

Thus ϕ fixes any vertex labeled with a two-letter word in the
elements of K.   By induction, ϕ then fixes every vertex
labeled with a finite word in the elements of K.   Therefore,
under the hypotheses, every ϕ that fixes vertex E fixes the
whole graph.   #

Sometimes, however, it is easier to show that some Cayley
graph cannot be 0-symmetric by applying the following theorem.

THEOREM 4.3:   A graph that is the Cayley graph of two non-
isomorphic groups cannot belong to class Z.

PROOF: Follows from (ii) above.   #

EXAMPLE: From the Cayley diagrams given in Figs. 3.1 and 3.2
it follows that the triangular prism cannot be 0-symmetric.
Indeed, it is 1-symmetric as noted in Section 1.

Part II   Graphs of Type ^1Z

5 GENERAL REMARKS ON GRAPHS OF TYPE ^1Z

From now on we shall use the letter R (with or without subscripts) exclusively for involutory generators; in the cases ^1Z, ^1S, and ^1T the letter S will be used for the generator of period $\geq$ 3. We can then state the following necessary condition for a graph to be ^1Z.

THEOREM 5.1: A group $H = \{R,S\}$ can have a 0-symmetric Cayley graph with respect to the generators R, S only if H admits no automorphism leaving R fixed and taking S into its inverse S^{-1}.

PROOF: Suppose that there were such an automorphism ϕ defined by

$$\phi(R) = R, \quad \phi(S) = S^{-1} . \tag{5.1}$$

The <u>same</u> mapping ϕ of the vertices of the Cayley graph will give us an automorphism of the graph that leaves E fixed (since ϕ is an automorphism of H), but interchanges the edges $\{E,S\}$ and $\{E,S^{-1}\}$. Hence the Cayley graph of H will not be 0-symmetric. #

REMARKS: (i) The existence of an automorphism ϕ such as that supposed in the proof of Theorem 5.1 does not mean that the Cayley graph is necessarily ^1T; it can be ^1S as shown by the example (already mentioned in Section 3) of K_4, which is the Cayley graph of the group C_4 when defined by the following relations:

$$S^2 = R, \quad R^2 = E . \tag{5.2}$$

23

(ii) That the necessary condition given in Theorem 5.1 is not sufficient can be seen from the following counterexample. Let H be the group generated by the permutations

$$R = (1\ 7)(2\ 6)(3\ 5),\quad S = (2\ 4\ 3\ 7\ 5\ 6)\ . \tag{5.3}$$

As an abstract group, H is isomorphic to the K-metacyclic group of order 42, since it might also be generated by S and

$$Q = RS^3 = (1\ 2\ 3\ 4\ 5\ 6\ 7)\ , \tag{5.4}$$

and the relations

$$Q^7 = S^6 = E,\quad S^{-1}QS = Q^3 \tag{5.5}$$

are just the defining relations of that well known group as can be seen, e.g., from Coxeter and Moser (1980), formula 1.89. (From these relations it can be seen that H can also be described as the holomorph of C_7.)

It follows from (5.3) that

$$RS^3RS^{-2}RS^2 = E, \tag{5.6}$$

but

$$RS^{-3}RS^2RS^{-2} = (1\ 6\ 4\ 2\ 7\ 5\ 3) \neq E, \tag{5.7}$$

showing that in this case (5.1) does not define an automorphism.

Because of $RS^{-2}RS^2 = (1\ 7\ 6\ 5\ 4\ 3\ 2)$ the relation

$$(RS^{-2}RS^2)^7 = E \tag{5.8}$$

gives us a hamiltonian circuit in the Cayley graph G, leading to the easily checked LCF code $[9,-9]^{21}$.

However, G is also a Cayley graph of the dihedral group D_{21} of order 42 when generated by the involutory generators R_1 and R_2 satisfying the relation

$$(R_1R_2)^{21} = E ,\tag{5.9}$$

and a third (of course redundant) involutory generator R_3 defined by

$$R_3 = (R_1R_2)^4 R_1 .\tag{5.10}$$

Indeed (5.9) gives us a hamiltonian circuit, and (5.10) shows us that the LCF code is again $[9,-9]^{21}$.

It follows now from Theorem 4.3 that G is not 0-symmetric. The same conclusion might also be reached by realizing that G is isomorphic to the graph of the hexagonal map $\{6,3\}_{4,1}$ on a torus, known to be 1-regular, as it has an automorphism group of order 126 (Coxeter and Moser 1980, p. 107).

Finally it might be remarked that we had defined the generators R and S of H by the permutations (5.3), but that an abstract definition would be available from (5.5), using (5.4). It is noteworthy however that two relations are sufficient as an abstract definition of H, namely (5.6) together with $R^2 = E$. Thus the K-metacyclic group of order 42 turns out to be the member $F^{3,-2,2} \cong F^{3,2,-2}$ of a family $F^{a,b,c}$ of groups defined by

$$RS^aRS^bRS^c = R^2 = E .\tag{5.11}$$

We shall encounter other members of that family below; it is not yet known which of the groups $F^{a,b,c}$ are of finite order, although many instances have been found by Campbell, et al. (1977). See also Coxeter and Frucht (1979).

THEOREM 6.1:   A graph of type ^{3}T or ^{3}Z cannot have girth 3.

PROOF: Suppose the existence of a graph G being ^{3}T or ^{3}Z and having girth 3, i.e., having triangles as subgraphs.   Any such triangle obviously would correspond to the relation

$$R_1 R_2 R_3 = E \; , \tag{6.1}$$

where the R_i (i=1, 2, 3) are the involutory generators of the group H whose Cayley graph is G.   But it follows from (6.1) that $R_3 = R_1 R_2$ is a redundant generator, and being involutory that $(R_1 R_2)^2 = E$.   Hence H would be the four-group $D_2 \cong C_2 \times C_2$, and G the complete graph K_4 which is ^{3}S (as already mentioned in Section 2).   #

Thus the only 0-symmetric graphs of girth 3 are of type ^{1}Z.   They can be obtained by a procedure that might be described as "blowing up the vertices of a 1-regular graph to triangles".   Indeed, in any trivalent graph G with, say, 2n vertices we can replace the vertices in an obvious way by triangles, thus obtaining again a trivalent graph, say tG, but with 6n vertices.   See, for example, the Cayley graph of the alternating group A_4 with respect to the generators R = (1 2)(3 4) and S = (1 2 3), corresponding to the Cayley diagram reproduced in Fig. 3.3a of Coxeter and Moser (1980). The same graph would result from blowing up the vertices of K_4 to triangles.*   If, as in this example, the original

---

*This is the procedure whereby the regular tetrahedron {3,3} is truncated to form the Archimedean "truncated tetrahedron" t{3,3}.

27

graph G belongs to class S, the derived graph tG is still
vertex-transitive, but either of class T or of class Z.
Indeed, an inspection of any s-arc of the symmetrical graph G
and its augmented arc in tG shows that if G is s-regular, the
derived graph tG is t-symmetric with t = s-1, and this holds
still in the simplest case s = 1, thus giving a 0-symmetric
graph tG of girth 3 when "blowing up" a 1-regular graph G.

    Vice versa, if we have a graph of type ^1z and girth 3, we
can "shrink" its triangles to points or simply consider the
triangles as vertices of a new graph that is easily seen to
be 1-regular. So we have proved the following.

THEOREM 6.2:   There is a one-one correspondence between the
1-regular graphs of class S and the 0-symmetric graphs of
girth 3, resulting from the replacement of vertices by triangles or vice versa. #

    Thus to obtain the 0-symmetric graphs of type ^1z and girth
3 with not more than 120 vertices we have to look only for 1-
regular graphs with not more than 40 vertices. There are
according to C.C. Sims (Boreham et al. 1974) only two. They
are the graphs arising from the hexagonal maps $\{6,3\}_{3,i}$
(i=1,2) on a torus, with 26 and 38 vertices respectively.
From their LCF codes $[7,-7]^{13}$ and $[15,-15]^{19}$ it is easy to
obtain also codes for our two 0-symmetric graphs:

$$[2,21,-2,2,-21,-2]^{13} \tag{6.2}$$

for $t\{6,3\}_{3,1}$ with 78 vertices and

$$[2,45,-2,2,-45,-2]^{19} \tag{6.3}$$

for $t\{6,3\}_{3,2}$ with 114 vertices. The relations giving the
hamiltonian circuits in the two graphs are

$$(S^{-2}RS^2R)^m = E \tag{6.4}$$

with m = 13 or 19. Both LCF codes have chord length number
N_c = 2. The graphs might also be described as Cayley graphs
for the groups $[6,3]^+_{3,i}$ (i=1,2).

REMARKS: (i) The two graphs just described are of course the
first two members of an infinite family. The next one would
be $t\{6,3\}_{4,1}$ with 126 vertices, obtained by "blowing up" the
graph $\{6,3\}_{4,1}$ with LCF code $[9,-9]^{21}$ that we had already
encountered in Section 5 as a Cayley graph of the K-metacyclic
group $F^{3,2,-2}$ of order 42.

   (ii) In group-theoretical terms the procedure of "blowing
up vertices to triangles" might be described as follows.

   Suppose that the given l-regular graph G is of type ^3S,
with three involutory generators R_i. Now consider the extension $\{S, R_1\}$ of the group $\{R_1,R_2,R_3\}$ defined by the relations

$$S^3 = E \text{ , and} \tag{6.5}$$

$$S^{-1}R_1S = R_2, \; S^{-1}R_2S = R_3, \; S^{-1}R_3S = R_1 . \tag{6.6}$$

Then tG will be the Cayley graph of $\{S, R_1\}$. The details are
left as an exercise to the reader who might also check that
in the special case of the two graphs with less than 120
vertices the corresponding groups can be defined by the following relations:

$$S^3 = R^2 = (SR)^6 = (S^{-1}RSR)^3(SRS^{-1}R)^i = E \text{ (i=1,2) .} \tag{6.7}$$

7   THE GROUPS Z(m,n,k)

A main source of graphs of type ^{1}Z and girth at least
equal to 4 is a family of groups that will be called Z(m,n,k);
it may be described as follows.

It is known (Coxeter and Moser 1980, (1.83) and (1.84))
that a group of order ms is defined by

$$Q^m = S^S = E, \quad S^{-1}QS = Q^k \qquad (1 \leq k < m) \tag{7.1}$$

where, in order for these relations to be consistent, k must
satisfy

$$k^S \equiv 1 \pmod{m} . \tag{7.2}$$

A special case is the K-metacyclic group of order p(p-1)
arising from (7.1) when m = p is a prime, s = p - 1, and k is
a primitive root (mod p). The instance p = 7 has been encountered before in (5.5), where we saw that the same group
could be generated by one involutory and one non-involutory
element, thus leading to a trivalent Cayley graph. The same
is true not only for the K-metacyclic groups of order p(p-1)
in general (p being any odd prime), but more generally whenever s in (7.1) is even, say

$$s = 2n , \tag{7.3}$$

and when the exponent k satisfies, instead of (7.2), the
stronger condition

$$k^n \equiv -1 \pmod{m} . \tag{7.4}$$

Indeed, we shall prove the following.

THEOREM 7.1:   If (7.3) and (7.4) are fulfilled, then the group
(7.1) of order 2mn has a trivalent Cayley graph with respect
to {R, S}, where the involutory generator R is defined by

$$R = QS^n . \tag{7.5}$$

PROOF: Because of (7.1) and (7.3) we have $S^{2n} = E$, and so
(7.5) is equivalent to

$$Q = RS^n , \tag{7.6}$$

thus showing that R and S generate the same group (7.1) as Q
and S.   It remains only to be shown that R is involutory; but
it is easily seen that

$$R^2 = QS^n \cdot QS^n = Q \cdot S^{-n}QS^n = Q^{1+k^n} = E , \tag{7.7}$$

where we have used (7.4) and the case i = n of the relation

$$S^{-i}QS^i = Q^{k^i} , \tag{7.8}$$

which is a consequence of (7.1).   #

In terms of the generators R and S, the defining relations of our group of order 2mn will be

$$R^2 = S^{2n} = (RS^n)^m = E, \ S^{-1}RS^{n+1} = (RS^n)^k , \tag{7.9}$$

with k satisfying (7.4).   We shall call this group $Z(m,n,k)$.
For instance, the K-metacyclic group of order p(p-1) will be
$Z(p, \frac{p-1}{2}, k)$, k being a primitive root (mod p).

Since we are mainly interested in 0-symmetric graphs, we
have still to ask when these groups $Z(m,n,k)$ satisfy the
necessary condition of Theorem 5.1.   To this purpose let us

suppose the existence of an automorphism (5.1) leaving R fixed
and taking S into its inverse S^{-1}. Such an automorphism would
transform the last of the relations (7.9) into

$$SRS^{n-1} = (RS^n)^k \tag{7.10}$$

(because of $S^{-n} = S^n$), and it would follow from (7.9) and
(7.10) that

$$RS^n = S^{-1}(RS^n)^k S = (S^{-1}RS^{n+1})^k = \{(RS^n)^k\}^k = (RS^n)^{k^2}.$$

$$\tag{7.11}$$

Hence the existence of an automorphism (5.1) is seen to be
equivalent to the congruence

$$k^2 \equiv 1 \pmod{m}, \tag{7.12}$$

and we have obtained the following.

THEOREM 7.2: The group $Z(m,n,k)$ defined by (7.9) has with
respect to the generators R and S a trivalent Cayley graph
that cannot be 0-symmetric if k satisfies both the conditions
(7.4) and (7.12).   #

Hence to obtain Cayley graphs that are possibly 0-
symmetric we have to consider only those groups (7.9) where k
satisfies

$$\left.\begin{array}{l} k^n \equiv -1 \\ \\ k^2 \not\equiv 1 \end{array}\right\} \pmod{m} . \tag{7.13}$$

(We say "possibly" because the second condition in (7.13) is
only necessary, but not sufficient, as can be seen from the
discussion of the instance $Z(7,3,3)$ in Section 5.)

REMARKS: (i) If n is even, (7.12) is obviously incompatible with (7.4); so the second condition in (7.13) is really needed only if n is odd.

(ii) By (7.13) the values k = 1 and k = m - 1 are excluded; so we shall suppose from now on that

$$2 \leq k \leq m-2 \ . \tag{7.14}$$

(iii) For given values of m and n, the number of integers k satisfying (7.13) and (7.14) is even. (Don't forget that 0 is an even number!)

Indeed it is known that to any k satisfying (7.14) there exists an inverse (mod m), say k', such that

$$kk' \equiv 1 \pmod{m} \ , \tag{7.15}$$

and satisfying (7.14) too. Then it is easily checked that if k satisfies (7.13), so does k'. Finally it has to be shown that

$$k' \neq k; \tag{7.16}$$

but under the assumption that k = k', (7.15) would contradict the second condition in (7.13).

The groups Z(m,n,k) and Z(m,n,k') corresponding to such a pair of inverses (mod m) are however not only isomorphic (as abstract groups), but also have isomorphic Cayley graphs with respect to the generators defined in (7.9). This is due to the fact that replacing k in (7.9) by its inverse k' (mod m) is equivalent to replacing the generators {R,S} by {R,S^{-1}}. Hence only the arrows of the S-edges in the Cayley diagram are inverted, but such a change doesn't affect the Cayley graph.

(iv) We shall not consider here in its generality the problem of the possible isomorphism between two groups $Z(m,n,k)$ and $Z(m,n,\kappa)$ if k and κ are not inverses (mod m). Because of our limitation to groups at most of order 120 the only instance where this problem might arise is that of the groups $Z(11,5,2) \cong Z(11,5,6)$ and $Z(11,5,7) \cong Z(11,5,8)$, but in this case it is no problem at all: 2,6,7 and 8 are the primitive roots (mod 11) and so all four of the groups are isomorphic to the K-metacyclic group of order 110. In the notation of (5.11), the first and second are $F^{4,3,-2}$ whereas the third and fourth are $F^{4,2,-1}$. The isomorphism $F^{4,3,-2} \cong F^{4,2,-1}$ can be checked by writing S^3 for S in the former. (See Coxeter and Frucht 1979.)

8   GRAPHS OF TYPE ^1z AND GIRTH 4 THAT ARE CAYLEY GRAPHS OF GROUPS $Z(m,2,k)$

In this section we consider the Cayley graphs of the groups $Z(m,2,k)$ of order $4m$, that is, the special case

$$n = 2 \tag{8.1}$$

of the groups $Z(m,n,k)$ defined by (7.9).   In this case we have the defining relations

$$R^2 = S^4 = (RS^2)^m = E, \quad S^{-1}RS^3 = (RS^2)^k \ , \tag{8.2}$$

where the conditions (7.13) now reduce to

$$k^2 \equiv -1 \pmod{m} \ . \tag{8.3}$$

It follows from (8.3) that the inverse of k (mod m) is now

$$k' = m - k \ . \tag{8.4}$$

Thus, according to Remark (iii) after Theorem 7.2, two values of k whose sum equals m give rise to isomorphic Cayley graphs. Both values of k are given in the following Table 8.1, which

TABLE 8.1

No.	Order $4m$	m	k	b	c
1	20	5	2,3	2	1
2	40	10	3,7	3	1
3	52	13	5,8	3	2
4	68	17	4,13	4	1
5	100	25	7,18	4	3
6	104	26	5,21	5	1
7	116	29	12,17	5	2

contains the seven groups $Z(m,2,k)$ of order at most 120.
(The meaning of the last two columns will be explained later
in this section.)

The Cayley graphs of these groups have girth $g = 4$,
because of the squares corresponding to the first of the
relations (8.2). There exist also octagons, because of the
relation

$$(SR)^4 = E \qquad\qquad (8.5)$$

that can be shown to be a consequence of (8.2); the proof is
left to the reader as an exercise. It is easier to check
(8.5) in the Cayley diagram of the groups, which can be
drawn as a tessellation like that sketched in Fig. 8.1.

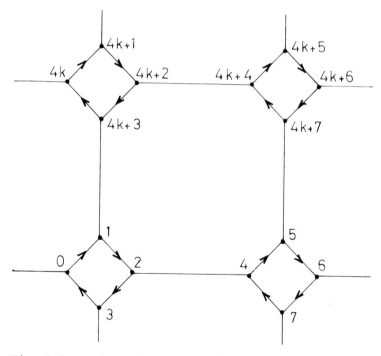

Fig. 8.1 Cayley diagram of $Z(m,2,k)$. Edges with arrowhead correspond to generator S, those without
to R; $4t + u$ means $(S^2R)^t S^u$, $t = 0, 1, \ldots, m-1$,
$u = 0, 1, 2, 3$.

Instead of (8.2), other relations can be used for defining the groups $Z(m,2,k)$, e.g.,

$$S^4 = R^2 = (SR)^4 = (SRS)^b \cdot (S^2R)^c = E \; , \qquad (8.6)$$

where

$$b > c > 0 \; , \qquad (8.7)$$

a possibility that has been considered by Coxeter (1975). By
analogy with Section 6, we may call this graph $t\{4,4\}_{b,c}$.
The relation between the parameters m and k used in (8.2) and
the parameters b and c used in (8.6) is

$$m = b^2 + c^2 \; , \qquad (8.8)$$

while one of the two possible values of k is the solution of
the congruence

$$bk + c \equiv 0 \pmod{m} \; . \qquad (8.9)$$

(See Table 8.1 above.)

It should be pointed out, however, that such a solution
may fail to exist; within the range of values considered
in this study ($m \le 30$) this happens only once, namely if

$$b = 4, \quad c = 2 \; . \qquad (8.10)$$

The group $[4,4]_{4,2}^+$ of order 80 then defined by (8.6) still has
a 0-symmetric Cayley graph (which will be considered at the
end of Section 11) but it is not a group $Z(20,2,k)$ at all.

Returning to the Cayley graphs of the groups $Z(m,2,k)$
and recalling the convenience of describing them by the LCF

code explained in Section 3, we should look in the first place for a "nice" hamiltonian circuit. For this purpose let us consider separately the cases of odd and even values of m.

It can be seen from the Cayley diagram (Fig. 8.1) that the word $(SRS^{-1}R)^i$ corresponds to a path leading, for instance, from the vertex "0" to that labeled $4i(k+1)$, and this is an open path for

$$i = 1, 2, 3, \ldots, \frac{m}{(k+1,m)} - 1 . \tag{8.11}$$

It becomes a closed circuit for

$$i = \frac{m}{(k+1,m)} . \tag{8.12}$$

Hence, if k and m are such that

$$(k+1, m) = 1 \tag{8.13}$$

—as is the case for Nos. 1, 3, 4, 5 and 7 of Table 8.1, corresponding to odd values of m—then the relation

$$(SRS^{-1}R)^m = E \tag{8.14}$$

gives us a hamiltonian circuit. It is easily checked that the LCF code for this hamiltonian circuit has "chord length number" $N_c = 1$ and is given by

$$[2k, 2k, -2k, -2k]^m . \tag{8.15}$$

Here, of the two values of k given in Table 8.1, the odd one always has to be taken (if k were even, (8.15) would not be an LCF code at all!). So we have for the five graphs with odd m the LCF notations given in Table 8.2. All five are 0-symmetric.

TABLE 8.2

No.	Number of vertices	k(odd)	LCF code
1	20	3	$[6, 6, -6, -6]^5$
3	52	5	$[10, 10, -10, -10]^{13}$
4	68	13	$[26, 26, -26, -26]^{17}$
5	100	7	$[14, 14, -14, -14]^{25}$
7	116	17	$[34, 34, -34, -34]^{29}$

It might be remarked that the graph No. 1 is the smallest known instance of a graph of type ^1Z; its group, the K-metacyclic group for p = 5, might also be described as $F^{2,1,-1}$ since the permutations R = (2 3)(4 5), S = (1 2 3 4) satisfy (5.11) with a = 2, b = 1, c = -1.

Passing to the case of m even, we remark that it follows from (8.3) that both values of k must be odd; hence (k+1, m) is at least 2 and the relation (8.13) does not yield a hamiltonian circuit. In the two cases of interest here, Nos. 2 and 6 of Table 8.1, it is, however, easy to find a hamiltonian circuit. In the case of the graph No. 2 with 40 vertices the relation

$$(S^{-3}RS^3R)^5 = E \qquad\qquad (8.16)$$

gives us a hamiltonian circuit with "chord length number" N_c = 3 and LCF code $[3,13,19,-3;-]^5$ (see Section 3 for notation). In an analogous fashion we can use for the graph No. 6 with 104 vertices the relation

$$(S^3RS^{-3}R)^{13} = E , \qquad\qquad (8.17)$$

thus obtaining the LCF code $[3,37,43,-3;-]^{13}$.

It is easily checked that both graphs are 0-symmetric. As to their groups, it might be pointed out that they are direct products according to the general relation

$$Z(2m,2,k) \cong C_2 \times Z(m,2,k) \ , \tag{8.18}$$

which holds whenever

$$m \equiv k \equiv 1 \ (\text{mod } 2) \tag{8.19}$$

and is easily proved. (More on the possibility of obtaining 0-symmetric graphs by using direct products of groups will be said below in Section 11.)

9 GRAPHS OF TYPE ^{1}Z THAT ARE CAYLEY GRAPHS OF GROUPS Z(m,n,k), n > 2

We now return to the groups Z(m,n,k) defined by (7.9), confining ourselves to those satisfying the condition (7.13) in order to obtain possibly 0-symmetric graphs, and we suppose in this section that

$$n > 2 \ . \tag{9.1}$$

There are ten such groups of order $2mn \leq 120$; their numerical properties are given in Table 9.1. The two values of k given for each group are inverses (mod m), and so they yield isomorphic Cayley graphs (see Remark (iii) after Theorem 7.2).

Since the relation $S^{2n} = E$ implies the existence of (2n)-gons in the Cayley graph, the girth g of the graph satisfies the inequality

$$g \leq 2n \ ; \tag{9.2}$$

TABLE 9.1

No.	Order 2mn	m	n	k	g	0-symmetric?
1	42	7	3	3,5	6	no
2	54	9	3	2,5	6	yes
3	60	5	6	2,3	9	yes
4	78	13	3	4,10	6	no
5	84	14	3	3,5	6	yes
6	100	5	10	2,3	10	yes
7	108	18	3	5,11	6	yes
8A	110	11	5	2,6	10	yes
8B	110	11	5	7,8	10	no
9	114	19	3	8,12	6	no
10	120	10	6	3,7	10	yes

since, however, g can be smaller than 2n, we have included in Table 9.1 a column with the value of g.

Comments on the Groups of Table 9.1 and their Cayley Graphs

No. 1:   This group $F^{3,2,-2}$ (the K-metacyclic group for p = 7) and its Cayley graph (of type ^{1}S) have already been discussed in Section 5.

No. 2:   From the representation of this group by permutations:

$$R = (1\ 9)(2\ 8)(3\ 7)(4\ 6),\ S = (2\ 3\ 5\ 9\ 8\ 6)(4\ 7) \qquad (9.3)$$

it can be readily seen that

$$Z(9,3,2) \cong F^{3,-1,1} \cong F^{3,1,-1}\ , \qquad (9.4)$$

in the notation of (5.11), and that the relation

$$(S^5 R)^9 = E \qquad (9.5)$$

gives a hamiltonian circuit with "chord length number" $N_c = 3$ and the (not antipalindromic) LCF code:

$$[5,-11,11,25,-25,-5]^9\ . \qquad (9.6)$$

No. 3:   From the representation of this group by permutations

$$R = (1\ 2)(3\ 4),\ S = (2\ 3\ 4\ 5)(6\ 7\ 8) \qquad (9.7)$$

we see that

$$Z(5,6,2) \cong F^{3,1,2} \cong F^{3,2,1}\ . \qquad (9.8)$$

Since R and S^3 generate $F^{2,1,-1}$, and $S^4 = (6\ 7\ 8)$, we have

$$F^{3,2,1} \cong C_3 \times F^{2,1,-1}\ ; \qquad (9.9)$$

that is, $Z(5,6,2) \cong C_3 \times Z(5,2,2)$.   The relation

$$(S^5RS^{-5}R)^5 = E \qquad (9.10)$$

gives a hamiltonian circuit with $N_c = 4$ and the LCF code

$$[18,9,-17,29,-9,18;-]^5 . \qquad (9.11)$$

Since (9.8) is equivalent to the relation

$$RS^3RSRS^2 = E \qquad (9.12)$$

which produces 9-gons in the graph, and since no 'shorter' relation (apart from $R^2 = E$) occurs in the group, the graph has girth 9.   In fact, each vertex belongs to nine 9-gons; e.g., the vertex E belongs to the 9-gons representing the cyclic permutations of the word (9.12):

$$RS^3RSRS^2, \ S^3RSRS^2R, \ S^2RSRS^2RS, \ldots, SRS^3RSRS .$$

Recently, Biggs and Hoare (1980) discovered a trivalent graph of girth 9 with only 58 vertices.   Since their graph is not vertex-transitive, it is likely that the graph (9.11) is the smallest 0-symmetric graph of girth 9.   (See also Frucht 1955.)

No. 4:   This group can be represented by the following matrices mod 13:

$$R \equiv \begin{pmatrix} -1 & 0 \\ -1 & 1 \end{pmatrix} , \quad S \equiv \begin{pmatrix} 4 & 0 \\ 0 & 1 \end{pmatrix} \quad (\text{mod } 13) , \qquad (9.13)$$

and from this representation it is easily seen that the relation

$$(RS^2RS^{-2})^{13} = E \qquad (9.14)$$

gives a hamiltonian circuit with LCF code $[33,-33]^{39}$. It follows then from Theorem 4.3 that this graph is not 0-symmetric since it is also a Cayley graph for the dihedral group D_{39} of the same order 78, with respect to appropriate involutory generators satisfying the relations

$$(R_1 R_2)^{39} = E, \quad R_3 = (R_1 R_2)^{16} R_1 \tag{9.15}$$

(see (5.9) and (5.10) for a similar case encountered above).

No. 5: This group allows the representation by permutations

$$R = (1\ 7)(2\ 6)(3\ 5)(8\ 9), \quad S = (2\ 4\ 3\ 7\ 5\ 6) . \tag{9.16}$$

Comparison with the analogous representation (5.3) of the group No. 1 of Table 9.1 shows that

$$Z(14,3,3) \cong C_2 \times Z(7,3,3) \cong C_2 \times F^{3,2,-2} ; \tag{9.17}$$

but contrary to what happened in the former case, we now obtain a 0-symmetric Cayley graph with the LCF code

$$[5,21,-29,-19,39,-5;-]^7 , \tag{9.18}$$

using the hamiltonian circuit coming from the relation

$$(S^5 R S^{-5} R)^7 = E . \tag{9.19}$$

No. 6: Like No. 3—see (9.9)—this group also is a direct product, namely

$$Z(5,10,2) \cong C_5 \times Z(5,2,2) \cong C_5 \times F^{2,1,-1} , \tag{9.20}$$

as can be seen from this representation by permutations, analogous to (9.7):

$$R = (1\ 2)(3\ 4), \quad S = (2\ 3\ 4\ 5)(6\ 7\ 8\ 9\ 10) . \tag{9.21}$$

No "nicer" hamiltonian circuit seems to exist in the graph than that corresponding to relations like

$$(S^9RS^{-9}R)^5 = E , \qquad (9.22)$$

yielding the rather lengthy LCF code:

$$[30,17,-25,-47,-9,9,-33,45,-17,30;-]^5 . \qquad (9.23)$$

It follows from (3.8) and (9.23) that the girth of the graph cannot be higher than 10, and from a drawing it can be seen that it is indeed 10.

No. 7:   As can be seen from the representation by permutations

$$R = (1\ 9)(2\ 8)(3\ 7)(4\ 6)(10\ 11), \ S = (2\ 6\ 8\ 9\ 5\ 3)(\ 4\ 7) ,$$
$$(9.24)$$

this group is the direct product of No. 2 and a group of order 2:

$$Z(18,3,5) \cong C_2 \times Z(9,3,2) \cong C_2 \times F^{3,1,-1} . \qquad (9.25)$$

The relation

$$(S^5R)^{18} = E \qquad (9.26)$$

gives a hamiltonian circuit with the non-antipalindromic LCF code

$$[5,-29,29,43,-43,-5]^{18} . \qquad (9.27)$$

No. 8:   It has already been mentioned—see Remark (iv) at the end of Section 7—that the groups $Z(11,5,2)$ and $Z(11,5,8)$ are abstractly isomorphic (both being isomorphic to the

K-metacyclic group for p = 11); but the Cayley graphs 8A and
8B turn out to be different because of the difference in
generators.

Both graphs have girth 10 and a hamiltonian circuit coming
from the relation

$$(S^4RS^{-4}R)^{11} = E \ , \tag{9.28}$$

with slightly different LCF codes, namely

$$[35,17,25,-27,35;-]^{11} \tag{9.29}$$

for 8A and

$$[45,27,15,53,45;-]^{11} \tag{9.30}$$

for 8B, as can be checked by using the following representation:

$$R = (1 \ 11)(2 \ 10)(3 \ 9)(4 \ 8)(5 \ 7) \ , \tag{9.31}$$

$$S = \begin{cases} (2 \ 3 \ 5 \ 9 \ 6 \ 11 \ 10 \ 8 \ 4 \ 7) \text{ for 8A, which is } F^{4,3,-2} \\ (2 \ 8 \ 6 \ 3 \ 4 \ 11 \ 5 \ 7 \ 10 \ 9) \text{ for 8B, which is } F^{4,2,-1} \end{cases} . \tag{9.32}$$

That the two graphs are not isomorphic can be seen, for
example, from the fact that in 8A there are 36 vertices at
distance 5 from any vertex, but in 8B only 30.

Still more surprising is the fact that only the graph 8A
is 0-symmetric; the graph 8B is of type ^1S, as has been shown
by Coxeter and Frucht (1979).

No. 9: We omit details since this case is analogous to No. 4.
The relation

$$(RS^{-2}RS^2)^{19} = E \tag{9.33}$$

gives a hamiltonian circuit with the LCF code $[15,-15]^{57}$, which corresponds also to a Cayley graph of the dihedral group D_{57}, thus showing that the graph is not 0-symmetric.

No. 10:   Comparing the representation

$$R = (1\ 2)(3\ 4)(6\ 7),\ S = (2\ 5\ 4\ 3)(8\ 9\ 10) \tag{9.34}$$

of $Z(10,6,3)$ with (9.7), it follows from (9.9) that

$$Z(10,6,3) \cong C_2 \times C_3 \times Z(5,2,2) \cong C_6 \times F^{2,1,-1} \tag{9.35}$$

Because of the relation

$$S^4 RS^{-4}R = E \tag{9.36}$$

the Cayley graph has girth 10.   No "nicer" hamiltonian circuit has been found than that implied by the relation

$$(S^{11}RS^{-11}R)^5 = E\ , \tag{9.37}$$

with the rather lengthy LCF code:

$$[11,-51,-29,17,-9,-59,-37,9,-17,53,-45,-11;-]^5\ . \tag{9.38}$$

10 MORE 0-SYMMETRIC CAYLEY GRAPHS OF $Z(m,n,k)$ OBTAINED BY CHANGE OF GENERATORS

In the groups $Z(m,2,k)$ considered in Section 8, the elements S and SR have the same period (namely 4) as can be seen from (8.2) and (8.5). However, in the groups $Z(m,n,k)$ with $n > 2$ (considered in Section 9) it can happen that the period of SR is different from that of S, i.e. from 2n, according to (7.9). In these cases we can obtain a second trivalent Cayley graph for the same group $Z(m,n,k)$, by using R and P = SR as generators.

Maintaining the numeration of Table 9.1 we present in Table 10.1 a list of those cases where the periods of S and SR are different.

TABLE 10.1

No.	Order of group	Period of S	P=SR	g	0-symmetric?
1	42	6	3	3	no
2	54	6	9	8	yes
4	78	6	3	3	yes
7	108	6	18	8	yes
8A,8B	110	10	5	5	yes
9	114	6	3	3	yes

Comments on the Graphs Listed in Table 10.1

<u>No. 1</u>: It follows from Theorem 6.2 that this graph is ^{1}T; indeed by shrinking its triangles to vertices the graph reduces to the well known 6-cage (also called the Heawood graph; see Harary 1969, Fig. 14.11) with LCF code [5,-5]^7, which is

48

4-regular. Hence the graph No. 1 is more precisely 3-symmetric. In the notation of Section 6, it is $t\{6,3\}_{2,1}$.

<u>No. 2</u>: From the representation of the group $F^{3,1,-1}$ by the permutations

$$P = (1\ 9\ 2\ 7\ 6\ 8\ 4\ 3\ 5),\ R = (1\ 9)(2\ 8)(3\ 7)(4\ 6)\qquad (10.1)$$

it follows readily that the relation

$$(P^8R)^6 = E \qquad\qquad (10.2)$$

gives us a hamiltonian circuit with the (not antipalindromic) LCF code

$$[8,27,13,-7,-24,7,-13,24,-8]^6\ . \qquad\qquad (10.3)$$

The girth 8 is due to the existence of octagons coming from relations such as

$$(P^3R)^2 = E \ . \qquad\qquad (10.4)$$

<u>Nos. 4 and 9</u>: These are the graphs $t\{6,3\}_{3,1}$ and $t\{6,3\}_{3,2}$ described in Section 6.

<u>No. 7</u>: Using

$$P = (1\ 9\ 5\ 7\ 6\ 2\ 4\ 3\ 8)(10\ 11)$$
$$R = (1\ 9)(2\ 8)(3\ 7)(4\ 6)(10\ 11) \qquad (10.5)$$

we can easily check that the relation

$$(P^{17}R)^6 = E \qquad\qquad (10.6)$$

gives a hamiltonian circuit with the (not antipalindromic) LCF code

$$[17,-45,31,-7,-51,25,-13,51,19,-19,45,13,-25,-51,7,-31,51,-17]^6.$$
$$(10.7)$$

As in the case of No. 2, the girth 8 comes from relations like (10.4).

Nos. 8A and 8B: Both graphs have girth 5 (the only 0-symmetric graphs of girth 5 we have found with fewer than 120 vertices!).   The relation

$$(P^4RP^{-4}R)^{11} = E \tag{10.8}$$

gives a hamiltonian circuit with the following LCF codes:

$$[4,-53,25,-17,-4;-]^{11} \tag{10.9}$$

for 8A, and for 8B

$$[4,-43,15,33,-4;-]^{11}\ . \tag{10.10}$$

11 THE JUXTAPOSITION PROCEDURE

There exists a powerful method allowing us to derive from a 0-symmetric graph other (possibly) 0-symmetric graphs, but with a larger number of vertices; for this reason unfortunately only a few cases fall into the range of our study (not more than 120 vertices!). For graphs of type ^1z this "juxtaposition method" may be described as follows.

Let R,S be the generators of a group H with a 0-symmetric Cayley graph, and let T, U be the generators of another group K, about which we assume only that

$$T^2 = E .$$ (11.1)

(We do not exclude the possibility that T = E, or that U = E if T ≠ E. In both cases, of course, K would be a cyclic group.) Then define a group $\overline{H}$ with generators $\overline{R}$ and $\overline{S}$ by

$$\overline{R} = RT, \quad \overline{S} = SU ,$$ (11.2)

where R and S commute with T and U:

$$R, S \rightleftarrows T, U .$$ (11.3)

This juxtaposition procedure yields a group $\overline{H}$ that is isomorphic either to the direct product H × K or to some subgroup of this direct product—as will be shown by examples below. It can of course happen that $\overline{H} \cong H$, but, excluding such trivial cases, the Cayley graph of the larger group $\overline{H}$ with respect to the generators $\overline{R}$, $\overline{S}$ might again be 0-symmetric.

Thus a new 0-symmetric graph with 120 vertices and girth 4 might be obtained taking as H the group No. 1 of Table 8.1,

i.e., the K-metacyclic group $Z(5,2,2) \cong F^{2,-1,1}$ of order 20,
and as K the symmetric group S_3 of order 6, generating it
by two involutory elements. Then $\overline{H}$ will be the direct product $F^{2,-1,1} \times S_3$, as is easily checked using the permutations

$$\overline{R} = (1\ 2)(3\ 4)(6\ 8), \quad \overline{S} = (2\ 3\ 4\ 5)(6\ 7) \ . \tag{11.4}$$

The relation

$$(\overline{S}^3\overline{RS}^{-3}\overline{R})^{15} = E \tag{11.5}$$

gives a hamiltonian circuit with chord length number $N_c = 3$
and the LCF code

$$[3,53,19,-3;-]^{15} \ . \tag{11.6}$$

The same group $F^{2,-1,1} \times S_3$ gives rise to a second 0-
symmetric Cayley graph when we use, for $K \cong S_3$, one involutory generator and one of period 3; a representation by
permutations might be

$$R' = (1\ 2)(3\ 4)(6\ 8), \quad S' = (2\ 3\ 4\ 5)(6\ 7\ 8) \ . \tag{11.7}$$

The corresponding Cayley graph is of girth 8, the octagons
coming from relations like

$$(R'S')^4 = E \ . \tag{11.8}$$

We could only find the non-antipalindromic LCF code

$$[53,-33,-17,41,9,-53,53,39,7,17,-39,-53,$$
$$29,-9,-17,-7,33,-29,-19,39,-41,17,-39,19]^5 \tag{11.9}$$

coming from the relation

$$(VRV^{-1}SU^{-1}RUS^{-1})^5 = E \ , \tag{11.10}$$

where we have written

$$V = RS^2RS, \quad U = SRS^2R . \tag{11.11}$$

As mentioned above, $\bar{H}$ may be isomorphic to a subgroup of the direct product $H \times K$. This is precisely the case if we take as H again the K-metacyclic group $F^{2,-1,1}$ of order 20, but as K the dihedral group D_4, generated by two involutory elements. Thus $\bar{H}$ might be generated by the permutations

$$\bar{R} = (1\ 2)(3\ 4)(6\ 7)(8\ 9), \quad \bar{S} = (2\ 3\ 4\ 5)(7\ 9) . \tag{11.12}$$

It is easily checked that the group $\bar{H}$ is not isomorphic to the direct product $F^{2,-1,1} \times D_4$ (of order 160), but is only of order 80; indeed it contains only even permutations. The Cayley graph is of girth 4 and 0-symmetric; the relation

$$(\bar{S}^{-3}\overline{RS}^3\bar{R})^{10} = E \tag{11.13}$$

gives a hamiltonian circuit with chord length number $N_c = 3$ and LCF code:

$$[3,21,27,-3;-]^{10} . \tag{11.14}$$

The same graph has already been mentioned in Section 8; see the lines following (8.10).

12   THE CAYLEY GRAPHS OF THE GROUPS $F^{3,2,-1}$ AND $F^{4,2,1}$; THE EXTENDED LCF NOTATION

At the end of Section 5 we already mentioned the fact that $Z(7,3,3)$, the K-metacyclic group of order 42, can be defined by only two relations for its two generators; indeed that group can also be considered as the member $F^{3,2,-2}$ of a family $F^{a,b,c}$ of groups introduced by Campbell, et al. (1977) and defined by the two relations

$$R^2 = RS^aRS^bRS^c = E .$$

(12.1)

Four more such cases have been encountered above, namely;

(i)   No. 1 in Table 8.1: $Z(5,2,2) \cong F^{2,1,-1}$ or order 20;

(ii)   No. 2 in Table 9.1: $Z(9,3,2) \cong F^{3,1,-1}$ of order 54;

(iii)   No. 3 in Table 9.1: $Z(5,6,2) \cong F^{3,2,1}$ of order 60;

(iv)   No. 8A/8B in Table 9.1: $Z(11,5,2) \cong F^{4,2,1} \cong F^{4,3,-2}$ of order 110.

There are, however, groups belonging to the family $F^{a,b,c}$, but not to the family $Z(m,n,k)$ defined by (7.9), and two of them, $F^{3,2,-1}$ and $F^{4,2,1}$, will be considered in this section because they have 0-symmetric Cayley graphs with fewer than 120 vertices.

The Group $F^{3,2,-1}$ of Order 72

As generators of this group we can use the following permutations:

$$R = (1\ 3)(2\ 7)(4\ 9)(5\ 6), \quad S = (1\ 2\ 3\ 4\ 5\ 6\ 7\ 8),$$

(12.2)

which allow us to check that

$$(S^7R)^8 = E .$$

(12.3)

Of course, this relation yields only a 64-gon in the Cayley
graph (instead of a hamiltonian circuit, which we have not
been able to find), but this 64-gon can be used for a drawing
of the graph, as can be seen from Fig. 12.1. For the concise
description of the Cayley graph—which is easily seen to be 0-
symmetric and of girth 8, but not bipartite—the following
extended LCF notation might be used. We consider the 64-gon
corresponding to the relation (12.3) as if it were a
hamiltonian circuit and compute the chord lengths as usual;
but when we come to a vertex that is adjacent to a vertex of
the octagon not belonging to the 64-gon we insert in the LCF
code a fraction—in our example it will be $\infty/3$. Here the
numerator ∞ is conventional (telling us only that we have come
to a chord ending "at infinity": that is, outside the
"pseudohamiltonian" circuit), and the denominator 3 means that

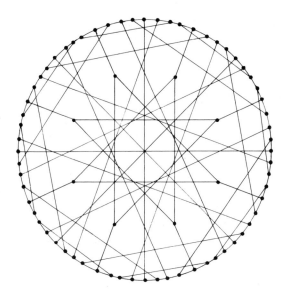

Fig. 12.1 Cayley graph of $F^{3,2,-1}$.

the octagon "at infinity" is not just a simple one (in which

case the denominator would be 1), but rather an octagram {8/3}

in which each vertex is adjacent to the third following one

(as in the interior polygon of a generalized Petersen graph).

From Fig. 12.1 it can now be seen that the Cayley graph of the

group $F^{3,2,-1}$ is described by the following extended LCF code:

$$[7,27,32,-14,-27,14,\infty/3,-7]^8 . \qquad\qquad (12.4)$$

REMARKS: (i) Also the Cayley graph for the group $F^{2,1,-1} \cong$

$Z(5,2,2)$ (No. 1 in Table 8.1), which in Table 8.2 was described

by the LCF code $[6,6,-6,-6]^5$, might be characterized by an

extended LCF notation, namely $[3,\infty/1,8,-3]^4$.

(ii)   An extremely concise description is offered by the

extended LCF notation for the Petersen graph and its general-

izations, viz. $[\infty/2]^5$ and, in general, $[\infty/k]^n$.

The Group $F^{4,2,1}$ of Order 112

Using as generators the permutations

$$\begin{aligned}
R &= (1\ 16)(2\ 11)(3\ 14)(4\ 9)(5\ 6)(7\ 10)(8\ 15)(12\ 13), \\
&\qquad\qquad\qquad\qquad\qquad\qquad\qquad\qquad\qquad\qquad (12.5) \\
S &= (1\ 2\ 3\ 4\ 5\ 6\ 7\ 8\ 9\ 10\ 11\ 12\ 13\ 14)(15\ 16),
\end{aligned}$$

one can easily check that this group has the following prop-

erties:

(i) S^7 is a central element;

(ii) a subgroup of index 2, say G_{56}, is generated by the

elements

$$R' = RS^7 \text{ and} \qquad\qquad\qquad\qquad\qquad (12.6)$$

$$Q = S^2; \qquad\qquad\qquad\qquad\qquad\qquad (12.7)$$

(iii) $F^{4,2,1}$ is isomorphic to the direct product $G_{56} \times C_2$;

(iv) a group isomorphic to G_{56} can be obtained as the homomorphic image $F^{4,2,1}/\{S^7\}$, i.e., using as defining relations

$$R^2 = RS^4RS^2RS = S^7 = E \ . \tag{12.8}$$

The Cayley graph of this group G_{56} will be studied below. Let us first consider those of the group $F^{4,2,1}$ (of order 112). According to different choices of the generators we have found the following three Cayley graphs, all of them being of type ^1Z.

(1) Using the generators R and S defined by (12.5) we obtain a bipartite 0-symmetric Cayley graph of girth 8 (Fig. 12.2), somewhat similar to that of the group $F^{3,2,-1}$ discussed before. Again we can obtain a concise description of the graph by using an extended LCF notation, namely

$$[-43,-25,\infty/1,-33,25,33,43]^{14} \ , \tag{12.9}$$

where the 98-gon used as a "pseudohamiltonian" circuit comes from the relation

$$(S^6R)^{14} = E \ . \tag{12.10}$$

(2) Replacing the involutory generator R with R' defined by (12.6) we obtain another 0-symmetric Cayley graph of girth 8; it is not bipartite (hence not isomorphic to the foregoing one) as can be seen from the existence of 9-gons corresponding to relations like

$$R'S^3R'S^{-1}R'S^{-2} = E \ , \tag{12.11}$$

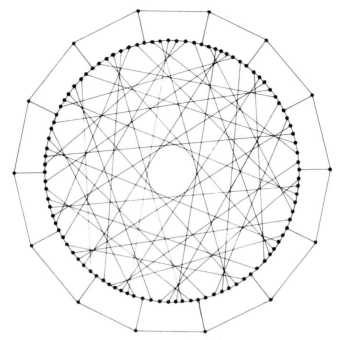

Fig. 12.2 Cayley graph of $F^{4,2,1}$.

or also from the non-antipalindromic LCF codes

$$[-30,-54,30,7,-27,-22,34,53,-34,27,-7,54,-53,-31,31,22]^7 ,$$

$$(12.12)$$

$$[34,-38,-34,7,37,10,-30,-43,30,-37,-7,38,43,-31,31,-10]^7 ,$$

$$(12.13)$$

coming from two hamiltonian circuits corresponding to the
relations

$$(S^4R'SR'S^{-1}R'SR'S^2R'S^\varepsilon R')^7 = E, \quad \varepsilon = \pm 1 . \qquad (12.14)$$

(3) The group $F^{4,2,1}$, defined by (12.6), can also be
generated by R and $Q = S^2$; indeed it follows from the definition of $F^{4,2,1}$ that

$$S = (RS^4RS^2R)^{-1} = (RQ^2RQR)^{-1} . \qquad (12.15)$$

Using R and Q as generators, we obtain a 0-symmetric Cayley graph of girth 7, the heptagons coming from the relation

$$Q^7 = E ; \qquad (12.16)$$

and a concise description of the graph is given by the extended LCF code

$$[6,24,\infty/6,-33,-24,33,-6]^{14} \qquad (12.17)$$

coming from the "pseudohamiltonian" 98-gon corresponding to the relation

$$(Q^6R)^{14} = E . \qquad (12.18)$$

The symbol $\infty/6$ in (12.17) indicates the occurrence of two heptagrams $\{7/3\}$.

The Subgroup G_{56} of $F^{4,2,1}$

It has already been mentioned above that the elements (12.6) and (12.7) generate a group of order 56; its Cayley graph is 0-symmetric and of girth 7. The relation

$$(Q^6R')^7 = E \qquad (12.19)$$

yields a "pseudohamiltonian" 49-gon with the extended LCF code

$$[6,24,\infty/1,16,-24,-16,-6]^7 . \qquad (12.20)$$

Having only 56 vertices this graph might well be the smallest 0-symmetric graph of girth 7.

Part III Graphs of Type ^3Z

13 THE PARAMETERS OF CAYLEY GRAPHS OF GROUPS WITH THREE INVOLUTORY GENERATORS

Before beginning the study of graphs of type ^3z (that is, of 0-symmetric Cayley graphs of groups with three involutory generators), it will be useful to introduce certain parameters that can be defined more generally for Cayley graphs with three involutory generators (and not only for the 0-symmetric graphs).

For that purpose let H be such a group with involutory generators R_1, R_2, R_3, and let

$$(R_1 R_2)^{P_3} = (R_2 R_3)^{P_1} = (R_3 R_1)^{P_2} = E , \qquad (13.1)$$

where it is understood that p_3, p_1, p_2 are the true periods of $R_1 R_2$, $R_2 R_3$, $R_3 R_1$ respectively (and not multiples of the periods). In general it will be tacitly assumed that the subindices of the three generators have been chosen in such a way that

$$p_3 \geq p_1 \geq p_2 . \qquad (13.2)$$

(However, for convenience, we sometimes violate this condition).

Now let q be the period of the product of the three generators; that is

$$(R_1 R_2 R_3)^q = E , \qquad (13.3)$$

it being again understood that q is the true period of $R_1 R_2 R_3$. Then it follows from $R_i^2 = E$ (i = 1,2,3) that q is

also the period of $R_1 R_3 R_2$, $R_2 R_1 R_3$, $R_2 R_3 R_1$, $R_3 R_1 R_2$ and $R_3 R_2 R_1$.
In other words, q is not affected by a permutation of the
generators.

The numbers p_3, p_1, p_2, q will be called the four _parameters_
of the Cayley graph of H corresponding to the generators R_i
(or also of the group H itself). As an instance let us consider the symmetric group S_4 of order 24: if we use as
generators the transpositions

$$R_1 = (1\ 2),\ R_2 = (1\ 3),\ R_3 = (1\ 4)\ , \qquad (13.4)$$

the parameters are 3,3,3,4; however, for the generators

$$R_1 = (1\ 2),\ R_2 = (2\ 3),\ R_3 = (3\ 4) \qquad (13.5)$$

we have the parameters 3,3,2,4. A third possibility—and the
only one leading to a 0-symmetric graph (which will be discussed below in Section 15 under No. 5)—is

$$R_1 = (1\ 4)(2\ 3),\ R_2 = (2\ 4),\ R_3 = (1\ 4)\ , \qquad (13.6)$$

with parameters 4,3,2,3.

It should however be pointed out that a Cayley graph of
type ^3S, ^3T or ^3Z is not always completely characterized by
its four parameters. For instance, the same parameters 3,3,3,4
just found for (13.4) are also those of the group with generators

$$R_1 = (1\ 2)(5\ 6),\ R_2 = (1\ 3)(5\ 7),\ R_3 = (1\ 4)(5\ 7), \qquad (13.7)$$

and this latter group, consisting of the even permutations in
the direct product $S_4 \times S_3$, is a subgroup of index 2 of that

direct product. Hence the Cayley graph of the group (13.7)

has 72 vertices—instead of 24 as had the Cayley graphs of S_4,

and in particular that of (13.4).

The reason why in general a Cayley graph of the kind we

are considering here is not completely defined by its four

parameters might be seen in the following.

The group with involutory generators R_i subject only to

the defining relations (13.1) and (13.3) is, in general, infi-

nite as can be seen, for example, from Coxeter (1970) where

it is called $((p_3, p_1, p_2; q))$. Hence in order to have a group

of finite order it may be necessary to add further relations

for the R_i, and such relations can be chosen in a rather arbi-

trary fashion, the only restriction being that the orders of

R_1R_2, R_2R_3, R_3R_1, and $R_1R_2R_3$ should not be lowered. Also,

in the few exceptional cases where the group $((p_3, p_1, p_2; q))$ is

of finite order, it may be still possible to obtain homomorphic

images of lower order with the same parameters by adding

further relations. The case 3,3,3,4 just considered is such

an exceptional case: the group $((3,3,3;4))$ is of order 72 as

can be seen from Table 3 of Coxeter (1970).*

Nevertheless, the parameters p_3, p_1, p_2, q prove to be useful

for describing or distinguishing Cayley graphs of groups with

three involutory generators. For the girth of such a graph,

for instance, we have the inequality

$$g \leq \min(2p_2, 3q) \tag{13.8}$$

---

*There is a misprint in Table 2 of Coxeter (1970): the group
of order 288 for the honeycomb $\{3,4,3\}_6$ is not $((3,3,3;4))$
but $((6,3,3;4))$.

as an immediate consequence of (13.1), (13.2), and (13.3).

In the example considered above, of the three possibilities for the group S_4, the only one leading to a 0-symmetric graph is ((4,3,2;3)); and so it would seem reasonable to hope that a Cayley graph be ^3Z whenever the parameters p_i satisfy the condition of "complete asymmetry":

$$(p_2-p_3)(p_3-p_1)(p_1-p_2) \neq 0 . \tag{13.9}$$

It turns out, however, that this condition is neither necessary nor sufficient.

That (13.9) is not* a necessary condition can be seen from counterexamples of graphs being 0-symmetric in spite of having the first three parameters equal. For instance, if the integer $n \geq 11$ satisfies

$$n \equiv \pm 1 \pmod{6} , \tag{13.10}$$

for the dihedral group D_n of order 2n we can choose the (redundant) generators

$$\left. \begin{array}{l} R_1 = (1\ 2)(3\ n)(4\ n-1) \qquad \cdots \qquad (m\ m+2) , \\[2mm] R_2 = (2\ n)(3\ n-1)(4\ n-2) \quad \cdots \quad (m\ m+1) , \\[2mm] R_3 = (1\ n-1)(2\ n-2)(3\ n-3) \cdots \quad (m-1\ m) , \end{array} \right\} \tag{13.11}$$

where $m = (n+1)/2$, thus obtaining a Cayley graph with parameters n,n,n,2, for which the relation

$$(R_1R_2)^n = E \tag{13.12}$$

---

*However, (13.9) is a necessary condition in special cases, for example, for the graphs considered below in Section 15.

gives a hamiltonian circuit with LCF code $[5,-5]^n$. According to Boreham, et al. (1974, p. 215) all the graphs with this code are 0-symmetric whenever $n \geq 9$.

That on the other hand (13.9) is not a sufficient condition for a Cayley graph to be ^3Z can be seen from the counter-example of the dihedral group $D_6 \cong ((6,3,2;2))$ with generators

$$R_1 = (1\ 2)(3\ 6)(4\ 5), \quad R_2 = (2\ 6)(3\ 5), \quad R_3 = (1\ 5)(2\ 4)$$

(13.13)

The resulting Cayley graph has a hamiltonian circuit coming from the relation

$$(R_1 R_2)^6 = E , \qquad\qquad (13.14)$$

and LCF code $[5,-5]^6$. Again according to Boreham, et al. (1974, p. 215), this graph is not ^3Z. (It is, of course, ^3T; no graph of class S with 12 vertices is known to exist.)

Instead of (13.9) there exists, however, the following theorem (quite analogous to Theorem 5.1 above) giving us a useful necessary condition for a Cayley graph to be ^3Z.

THEOREM 13.1: A group H with three involutory generators R_1, R_2, R_3 can have a 0-symmetric Cayley graph with respect to these generators only if H admits no non-identical automorphism permuting the three generators. #

We omit the proof, which is completely analogous to that of Theorem 5.1.

That the condition of the non-existence of a non-identical automorphism leaving the generators set-wise fixed is not sufficient for a Cayley graph to be ^3Z can be seen from the same counter-example (13.13) used for showing that (13.9) was

not such a sufficient condition either. (It is obvious that
in general (13.9) implies the non-existence of a non-identical
automorphism permuting the generators, but not vice versa.)

Finally it should be pointed out that also the juxtaposition procedure described in Section 11 for graphs of type ^1Z
can be extended to pairs of groups with three involutory
generators.

Indeed, let H and H' be two such groups, with generators
R_i and R_i', and with parameters p_3, p_1, p_2, q and p_3', p_1', p_2', q'
respectively (i = 1,2,3); then a group $\bar{H}$ with three involutory
generators $\bar{R}_i$ can be obtained by letting

$$\bar{R}_i = R_i R_i' \quad (i = 1,2,3) \tag{13.15}$$

and

$$R_i R_j' = R_j' R_i \quad (i,j = 1,2,3) \ . \tag{13.16}$$

As in the case considered in Section 11, $\bar{H}$ can be (but
need not be) isomorphic to the direct product H × H' (depending on some kind of "lack of affinity" between the two groups
that doesn't seem to have been studied so far). It is,
however, easy to give a rule for the parameters $\bar{p}_3, \bar{p}_1, \bar{p}_2, \bar{q}$ of
the Cayley graph of $\bar{H}$ with respect to the generators (13.15):
$\bar{p}_i$ is the least common multiple of p_i and p_i' (i = 1,2,3), and
$\bar{q}$ is the least common multiple of q and q'. (This follows
immediately from the definition of our parameters).

This rule, and the whole "juxtaposition procedure", remain
valid also in "degenerate" cases, where some of the R_i and/or
R_i' are not, strictly speaking, generators; what is needed is
only that they satisfy the relations

$$R_i^2 = R_i'^2 = E \quad (i = 1,2,3) \tag{13.17}$$

and that the $\bar{R}_i$ defined by (13.15) are really generators of a group $\bar{H}$, that is, all different, and different from E.

As an example let us consider the following choice for the "generators" of $H' \cong C_2$:

$$R_1' = R_2' = E, \; R_3' \neq E \; , \tag{13.18}$$

where R_3' commutes with R_1, R_2, and R_3. There are then only two possibilities: the new group $\bar{H}$ is either isomorphic to the direct product $H \times C_2$ (the interesting case!) or it is iso-morphic to H (a case without interest). It is also easy to see that the first case occurs only if in H there exists for the generators at least one relation

$$R_\alpha R_\beta R_\gamma \cdots R_\omega = E \tag{13.19}$$

in which R_3 appears an <u>odd</u> number of times as a factor. (E.g., $(R_1 R_3)^4 R_2 R_3 = E$ would be such a relation because R_3 appears five times.)

Indeed, if $\bar{H}$ were isomorphic to H it would follow from (13.19) that in $\bar{H}$ the analogous relation

$$\bar{R}_\alpha \bar{R}_\beta \bar{R}_\gamma \cdots \bar{R}_\omega = E \tag{13.20}$$

should hold. But because of (13.15) and (13.18) it would follow that

$$R_\alpha R_\beta R_\gamma \cdots R_\omega R_3'^\rho = E \; , \tag{13.21}$$

where ρ is odd, and because of (13.19) finally that

$$R_3'^\rho = E \; , \tag{13.22}$$

and hence $R_3' = E$, a contradiction to (13.18). So we have proved the following.

THEOREM 13.2: If a group H can be generated by three involutory elements, then the same holds for the direct product $H \times C_2$ if the presentation for H includes a relation in which one of the generators appears an odd number of times. #

COROLLARY 13.2.1: If the Cayley graph of a group H with three involutory generators is not bipartite or if at least one of the parameters p_3, p_1, p_2, q is odd, then also the direct product $H \times C_2$ has a presentation with three involutory generators. #

COROLLARY 13.2.2: If a group H can be generated by three involutory elements and if at least one of these belongs to the commutator subgroup of H, then also the direct product $H \times C_2$ can be generated by three involutory elements. #

14   THE COMPANION GRAPHS OF CAYLEY GRAPHS OF GIRTH 4

Let us now consider the special case of Cayley graphs of groups where two of the three involutory generators commute. In order to satisfy (13.2) it will be convenient to suppose that

$$R_1 R_3 = R_3 R_1 \; ; \tag{14.1}$$

indeed (14.1) is equivalent to

$$(R_1 R_3)^2 = E \; , \tag{14.2}$$

or, because of (13.1), to

$$p_2 = 2 \; . \tag{14.3}$$

Excluding the uninteresting case of the complete graph K_4 we know from Theorem 6.1 and its proof that Cayley graphs of groups with three involutory generators cannot have girth 3; so it follows from (13.8), or more directly from (14.2), that Cayley graphs satisfying (14.1) must have girth 4.

Since (14.2) tells us that the product $R_1 R_3$ is involutory, a convenient abbreviation for it will be R:

$$R = R_1 R_3 = R_3 R_1 \; . \tag{14.4}$$

Then it follows from

$$R_1 = R R_3 = R_3 R, \quad R_3 = R R_1 = R_1 R \tag{14.5}$$

that instead of R_1, R_2, R_3 we can use R, R_2, R_3, or also R_1, R_2, R, as generators for the same group H, thus obtaining two more Cayley graphs for H. We shall call them the companion graphs of the original graph. Since R commutes with R_1 and R_3, both companion graphs are again of girth 4. This follows also from the fact that replacing R_1 or R_3 by the new generator R means for the graph that we delete from each square corresponding to (14.2) one pair of opposite sides, replacing them by the diagonals. In other words, (14.3) remains valid also for the companion graphs.

We have mentioned in Section 13 that groups with three involutory generators (and parameters p_3, p_1, p_2, q) are either Coxeter's groups $((p_3, p_1, p_2; q))$ or homomorphic images of such groups. In the special case (14.3) that we are considering here, the group $((p, q, 2; r))$ has the presentation

$$(R_1 R_2)^p = (R_2 R_3)^q = (R_3 R_1)^2 = (R_1 R_2 R_3)^r = E \qquad (14.6)$$

(the R_i being of course involutory). The change of generators

$$R_1 = BC, \ R_2 = BCA, \ R_3 = CA;$$

$$A = R_1 R_2, \ B = R_2 R_3, \ C = R_3 R_2 R_1 \qquad (14.7)$$

yields the alternative presentation

$$A^p = B^q = C^r = (AB)^2 = (BC)^2 = (CA)^2 = (ABC)^2 = E \ , \quad (14.8)$$

in which the periods p,q,r can be freely permuted. In the notation of Coxeter (1970, pp. 11,29), we thus have

$$((p, q, 2; r)) \cong G^{p, q, r} \qquad (14.9)$$

and consequently

$$((q, r, 2; p)) \cong ((r, p, 2; q)) \cong ((p, q, 2; r)) \ . \qquad (14.10)$$

The corresponding graphs are clearly companions of one another in the above sense. Hence, for any group with parameters p,q,2,r, we can derive the parameters for the companion graphs by cyclically permuting p,q,r and then interchanging the first two, if necessary, so as to satisfy (13.2).

As an example let us consider the group

$$D_6 \cong G^{6,3,2} \cong ((3,2,2;6)) \cong ((6,2,2;3)) \cong ((6,3,2;2)) \tag{14.11}$$

for which the three Cayley diagrams are shown in Fig. 14.1. None of the corresponding graphs is 0-symmetric. They are respectively a hexagonal prism, $[6]^{12}$, and $[5,-5]^6$; for the last see also (13.13) and (13.14).

The inequalities $p \leq q \leq r$ used elsewhere for $G^{p,q,r}$ are, of course, unimportant, and we can feel free to change them as in the above description of D_6.

The groups $G^{p,q,r}$ that are of finite order not exceeding 120 and have 0-symmetric Cayley graphs of girth 4 will be considered in the next section. By Theorem 13.1 such a group cannot have a 0-symmetric graph if $p = q$. Thus only a few cases remain to be considered.

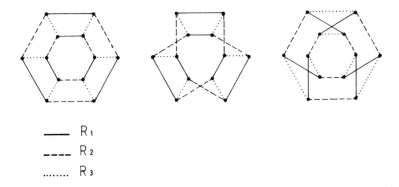

——— R₁
- - - - R₂
.... R₃

Fig. 14.1 Cayley diagram of D_6 and its companions.

To facilitate the discussion of the groups $G^{p,q,r}$ of order not exceeding 120 we have listed them in Table 15.1 in the same order in which they are given by Coxeter (1970, Table 1).

TABLE 15.1

No.	Group	Order
1	$G^{2,2s,2s} \cong D_{2s} \times C_2$	$8s$
2	$G^{2,s,2s} \cong D_s \times C_2$ (s odd)	$4s$
3	$G^{3,6,2n}$	$12n^2$
4	$G^{4,4,2n}$	$16n^2$
5	$G^{3,3,4} \cong S_4$	24
6	$G^{3,5,5} \cong A_5$	60
7	$G^{3,5,10} \cong A_5 \times C_2$	120

1.   The groups $G^{2,2s,2s}$ ($\cong D_{2s} \times C_2$) of order $8s$ cannot have 0-symmetric Cayley graphs because of the existence of an automorphism interchanging two of the involutory generators and leaving fixed the third.

2.   If s is odd, $G^{2,s,2s}$ is isomorphic to the dihedral group D_{2s} of order $4s$. As in the special case $s = 3$, already considered in Section 14 as an example, no 0-symmetric graphs arise.

3.   The next family to be considered is that of the groups $G^{3,6,2n}$ of order $12n^2$. Since $n = 1$ gives again the group $G^{2,3,6}$ just mentioned, we have to examine only the cases $n = 2$ and $n = 3$.

74

3a. $G^{3,4,6}$ (of order 48), or [3,4] in the notation of
Coxeter and Moser (1980), is isomorphic to the direct product
$S_4 \times C_2$.   Indeed, starting with the generators (13.6) of S_4,
the juxtaposition procedure explained at the end of Section
13 allows us to use

$$R_1 = (1\ 4)(2\ 3)(5\ 6),\ R_2 = (2\ 4)(5\ 6),\ R_3 = (1\ 4)(5\ 6)$$

$$(15.1)$$

as generators for the direct product $S_4 \times C_2$.   These generators satisfy the relations

$$(R_1 R_2)^4 = (R_2 R_3)^3 = (R_3 R_1)^2 = (R_1 R_2 R_3)^6 = E\ , \qquad (15.2)$$

which can be considered as defining relations of $G^{3,4,6}$, and,
since both groups have the same order 48, they are isomorphic.

The Cayley graph corresponding to the same generators, and
hence with parameters 4,3,2,6, is 0-symmetric and has the
remarkable feature of being planar; it is the skeleton of a
truncated cuboctahedron.   (See Fig. 15.1.)   Let us add the
information that the relation

$$(U_1 R_2 U_1 R_1)^4 = E\ , \qquad (15.3)$$

where $U_1 = R_3 R_2 R_1 R_2 R_3$, gives us a hamiltonian circuit with
the antipalindromic LCF code

$$[11,-3,7,5,-9,-11;-]^4 \qquad (15.4)$$

and the rather high chord length number $N_c = 5$.

This graph has two companion graphs (in the sense explained in Section 14) that turn out to be 0-symmetric too
(but not planar).   The first corresponds to the generators

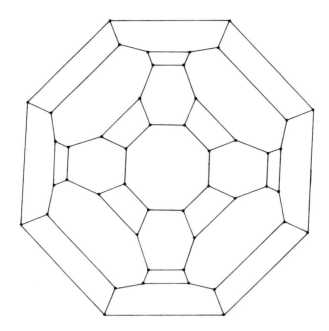

Fig. 15.1   Truncated cuboctahedron: Cayley graph of
$_G3,4,6$.

$R = (2\ 3),\ R_2 = (2\ 4)(5\ 6),\ R_3 = (1\ 4)(5\ 6)$ (15.5)

and has the parameters 6,3,2,4.   The relation

$(RR_3R_2RR_3RR_2R_3)^6 = E$ (15.6)

leads to a hamiltonian circuit with $N_c = 3$ and the LCF code

$[7,-3,3,17;-]^6$.   (15.7)

With a slight change of notation the generators of the
second companion graph can be taken as

$R_1' = (2\ 3),\ R_2' = (2\ 4)(5\ 6),\ R_3' = (1\ 4)(2\ 3)(5\ 6)$.
                (15.8)

The parameters are 6,4,2,3, and so this graph cannot be bipartite, since the relation $(R_1'R_2'R_3')^3 = E$ implies the existence of 9-gons. The relation

$$\{(R_3'R_2')^3 R_3'R_1'\}^6 = E \tag{15.9}$$

furnishes a hamiltonian circuit with $N_c = 3$ and the LCF code

$$[7,-3,-14,-14;-]^6 . \tag{15.10}$$

REMARK: Let

$$Z = (R_1R_2R_3)^3 = (5\ 6) ; \tag{15.11}$$

then the same abstract group $G^{3,4,6}$ can also be generated by the three involutory elements

$$R_1Z = (1\ 4)(2\ 3), \quad R_2 = (2\ 4)(5\ 6), \quad R_3Z = (1\ 4) . \tag{15.12}$$

The parameters are now 4,6,2,6, or rather 6,4,2,6 if we interchange two of the three generators. From a relation analogous to (15.9) we have then the LCF code

$$[7,-3,10,10;-]^6 , \tag{15.13}$$

which is quite similar to (15.10). The graph (15.13) is 0-symmetric, but not isomorphic to (15.10). We need not discuss here the companion graphs of (15.13); because of Theorem 13.1 no new 0-symmetric graphs arise.

Finally it might be remarked that the parameters 6,4,2,6 of (15.12) show that the group $G^{3,4,6}$ is abstractly a homomorphic image of the infinite group $G^{4,6,6}$ (Coxeter 1939, p. 119).

3b. For the group $G^{3,6,6}$ of order 108, Coxeter (1977, p. 262) gives the following involutory generators:

$$\left.\begin{array}{l} R_1 = (0\ 3)(4\ 7)(5\ 8)(b\ i)(c\ h)(d\ g)(e\ f)\ , \\[2mm] R_2 = (0\ 8)(2\ 3)(5\ 6)(a\ h)(b\ d)(e\ g)\ , \\[2mm] R_3 = (0\ 4)(1\ 6)(3\ 7)(b\ i)(c\ e)(f\ h)\ . \end{array}\right\} \qquad (15.14)$$

We have only interchanged R_1 and R_3 to have the parameters 6,3,2,6 for the 0-symmetric Cayley graph, for which a hamiltonian circuit can be found by using the relation

$$\{(U_1R_1)^2U_1R_2\}^6 = E\ , \qquad\qquad (15.15)$$

where U_1 again stands for $R_3R_2R_1R_2R_3$. The corresponding LCF notation for this graph is

$$[17,15,-5,-7,3,-11,-49,-3,-53;-]^6\ , \qquad\qquad (15.16)$$

with the rather high chord length number $N_c = 8$. We shall not discuss the companion graph with parameters 6,6,2,3, which cannot be 0-symmetric.

4. The family $G^{4,4,2n}$ of order $16n^2$ doesn't interest us here because it can give us possibly 0-symmetric Cayley graphs with parameters $2n,4,2,4$ only for $n \geq 3$.

5. The group $G^{3,3,4}$ (of order 24), or [3,3] in the notation of Coxeter and Moser (1980), is isomorphic to S_4. Since the parameters 3,3,2,4 would not serve, the only choice leading to a 0-symmetric graph—already mentioned above: see (13.6)— is 4,3,2,3. Using the representation

$$R_1 = (1\ 4)(2\ 3),\ R_2 = (2\ 4),\ R_3 = (1\ 4)\ , \qquad\qquad (15.17)$$

one can easily check that the relation

$$(R_3R_2R_3R_2R_3R_1)^4 = E \tag{15.18}$$

gives us a hamiltonian circuit with $N_c = 3$ and LCF code

$$[5,-3,12,12,3,-5]^4 \ . \tag{15.19}$$

The appearance here of the even number 12 shows that this graph is not bipartite. Thus it follows from Corollary 13.2.1 that the direct product $S_4 \times C_2$ can also be generated by three involutory elements. We have already used this fact; see 3a above in this section.

6.   The group $G^{3,5,5}$ (of order 60), or $[3,5]/C_2$ in the notation of Coxeter and Moser (1980) is isomorphic to the alternating group A_5.   The only convenient parameter combination for a 0-symmetric Cayley graph is 5,3,2,5, corresponding, for example, to the generators

$$R_1 = (1\ 2)(3\ 4),\quad R_2 = (2\ 3)(4\ 5),\quad R_3 = (1\ 4)(2\ 3)\ . \tag{15.20}$$

The relation

$$(U_1R_2U_1R_1)^5 = E\ , \tag{15.21}$$

where again $U_1 = R_3R_2R_1R_2R_3$, then gives us a hamiltonian circuit with $N_c = 5$ and LCF code

$$[11,-3,7,5,26,26;-]^5 \ . \tag{15.22}$$

Another LCF code, also with $N_c = 5$, namely

$$[11,9,-5,-7,26,26;-]^5 \ , \tag{15.23}$$

comes from the relation

$$(R_3R_2(R_1R_2)^4 R_3R_2)^5 = E .$$ (15.24)

The resulting graph is 0-symmetric, but not bipartite—a fact
that will allow us to obtain in the following case a presentation for the direct product $A_5 \times C_2$. Also Corollary 13.2.2
applies here.

7.   The group $G^{3,5,10} \cong A_5 \times C_2 \cong [3,5]$ is of order 120
and so is the last group in Table 1 of Coxeter (1970) that
lies within the range of our study.

The most interesting parameter combination is 5,3,2,10,
because the corresponding Cayley graph is planar: it is the
skeleton of a truncated icosidodecahedron (see Fig. 15.2), and

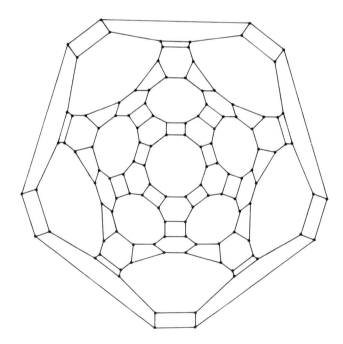

Fig. 15.2   Truncated icosidodecahedron: Cayley graph of
$G^{3,5,10}$.

is easily checked to be 0-symmetric.   Generators with these
parameters can be derived from (15.20) with the aid of the
juxtaposition procedure:

$$
\left.
\begin{array}{l}
R_1 \;=\; (1\ 2)(3\ 4)(6\ 7)\ , \\[2em]
R_2 \;=\; (2\ 3)(4\ 5)(6\ 7)\ , \\[2em]
R_3 \;=\; (1\ 4)(2\ 3)(6\ 7)\ .
\end{array}
\right\} \tag{15.25}
$$

These allow us to check that the relation

$$
(VR_1VR_2)^5 = E\ , \tag{15.26}
$$

where

$$
V = R_3R_2(R_1R_2)^2R_3R_2R_1R_2R_3\ , \tag{15.27}
$$

yields a hamiltonian circuit with the LCF code

$$
\begin{array}{l}
[23,21,-5,-7,-11,-13,13,11,-17,-19,3,-23, \\[0.5em]
23,-3,19,17,13,11,-11,-13,7,5,-21,-23]^5.
\end{array} \tag{15.28}
$$

It should be remarked that this code is not antipalindromic; however, the "chord lengths" d_i (defined in Section 3)
satisfy for $1 \le i \le 12$ the congruence

$$
d_{i+12} \equiv d_i \pmod{24}\ . \tag{15.29}
$$

The two companion graphs to be described at present are 0-
symmetric too, but not planar.   The first arises as the Cayley
graph of the same group, $G^{3,5,10}$, when using the generators

$$
R_1 = (1\ 3)(2\ 4),\ R_2 = (2\ 3)(4\ 5)(6\ 7)\ ,
$$
$$
R_3 = (1\ 4)(2\ 3)(6\ 7)\ . \tag{15.30}
$$

The parameters are now 10,3,2,5, and the relation

$$(R_1 R_2 R_3)^5 = E \qquad\qquad\qquad (15.31)$$

shows that this graph is not bipartite. This can also be seen from the LCF code

$$[11,21,19,55,53,26,26,-26,-26,-53,-55,-11;-]^5 , \qquad (15.32)$$

which is obtained from the relation

$$\{R_2 R_3 (R_2 R_1)^9 R_2 R_3 R_2 R_1\}^5 = E . \qquad\qquad (15.33)$$

The other companion graph corresponds to the generators

$$R_1 = (1\ 3)(2\ 4), \quad R_2 = (2\ 3)(4\ 5)(6\ 7) ,$$
$$\qquad\qquad\qquad\qquad\qquad\qquad\qquad\qquad (15.34)$$
$$R_3 = (1\ 2)(3\ 4)(6\ 7) .$$

and has parameters 10,5,2,3—hence it is likewise nonbipartite. It can be described by the LCF code of chord length number $N_c = 4$

$$[11,-3,-18,-18,3,25;-]^{10} , \qquad\qquad\qquad (15.35)$$

which corresponds to a hamiltonian circuit coming from the relation

$$\{R_1 (R_2 R_3)^2 R_1 (R_3 R_2)^2 R_1 R_3\}^{10} = E . \qquad\qquad (15.36)$$

For the same abstract group, $A_5 \times C_2$, it is possible to obtain a fourth 0-symmetric Cayley graph by using the same device as for $S_4 \times C_2$ (see the Remark after (15.10) above). Let

$$Z = (R_1 R_2 R_3)^5 = (6\ 7) ; \qquad\qquad\qquad (15.37)$$

then the generators

$$R_1Z = (1\ 2)(3\ 4)\ ,$$

$$R_2 = (2\ 3)(4\ 5)(6\ 7)\ ,$$

$$R_3Z = (1\ 4)(2\ 3)$$

$$(15.38)$$

yield a nonbipartite Cayley graph with parameters 10,6,2,10, and the LCF code

$$[59,21,19,-55,-57,36,36,-16,-16,-43,-45,-59;-]^5\ ,\quad (15.39)$$

coming from a relation analogous to (15.33).

REMARK: The graphs (15.1) and (15.25) are the only planar 0-symmetric graphs we have found in our study; we conjecture there are no others.

16 A PROCEDURE FOR OBTAINING BIPARTITE
CAYLEY GRAPHS OF GIRTH 4

Passing now to possibly 0-symmetric graphs that are Cayley graphs of <u>homomorphic images</u> of groups $G^{p,q,r}$ (and hence of girth 4), we shall confine our attention to those graphs which are also bipartite. In fact, it turns out that the non-bipartite graphs of girth 4 usually are companion graphs (in the sense of Section 14) of bipartite ones. The only exception we have found is the graph corresponding to the group $G^{3,5,5}$ (Section 15, item 6), where none of the three parameters p,q,r is even. On the other hand it is easy to obtain bipartite Cayley graphs of girth 4 using a procedure that will be described in this section.

As in Section 14 let us consider a group H with three involutory generators, two of which commute. Using the same notation as there, let

$$R = R_1 R_3 = R_3 R_1 \qquad\qquad (16.1)$$

be the involutory product of the two commuting generators; furthermore let

$$S = R_1 R_2 . \qquad\qquad (16.2)$$

To avoid trivial cases let us suppose that

$$p_3 > 2 ; \qquad\qquad (16.3)$$

then, according to the definition (13.1), S will not be involutory. Now let J be the subgroup of H generated by R and S; we show that J is of index 2 whenever the Cayley graph of H (with respect to the generators R_1, R_2, R_3) is bipartite.

84

In order to prove our assertion let us point out in the first place that J might also be defined as the subgroup generated by all the products $R_iR_k (i \neq k)$; this follows from (16.1), (16.2), and the fact that

$$R_3R_2 = RS, \quad R_2R_3 = S^{-1}R . \tag{16.4}$$

In other words, J contains those (and only those) elements of H that can be written as products of an <u>even</u> number of the generators R_1, R_2, R_3. Now suppose that such an element, say X, were at the same time a product of an <u>odd</u> number of the R_i:

$$X = U , \tag{16.5}$$

where U contains an odd number of generators as factors. But (16.5) would be equivalent to a relation of the form

$$UX^{-1} = E , \tag{16.6}$$

representing in the Cayley graph of H a closed walk with an <u>odd</u> number of edges. But such a walk cannot exist in a bipartite graph. Hence

$$H = J \cup R_1J , \tag{16.7}$$

where $R_1J = R_2J = R_3J$ is the left coset formed by those elements of H that are products of an odd number of generators; and from (16.7) it is obvious that J is indeed a subgroup of index 2.

Recalling that J was defined as being generated by the involutory element R and the non-involutory element S, we should point out that the trivalent Cayley graph of J (with respect to the generators R and S) cannot be 0-symmetric.

Indeed that graph does not fulfill the necessary condition of
Theorem 5.1 for a Cayley graph to be ^1Z, since J admits an
automorphism leaving R fixed and taking S into its inverse
S^{-1}, namely the automorphism defined by

$$\phi(X) = R_1^{-1}XR_1 = R_1XR_1 \text{ for all } X \in J , \qquad (16.8)$$

as is easily checked.

Why then our interest in the group J if its Cayley graph
is not 0-symmetric? The reason is that, inverting the direction of our reasoning, we might start with a convenient group
J whose Cayley graph is of type ^1T (or perhaps ^1S) and derive
from it another group H whose order is twice that of J, and
which has a Cayley graph that is possibly of type ^3Z. More
precisely our procedure might be described as follows.

Let J be any finite group generated by two elements R and
S satisfying

$$R^2 = E, \quad S^2 \neq E , \qquad (16.9)$$

and admitting an outer automorphism that leaves R fixed and
changes S into S^{-1}. To avoid the uninteresting case of
dihedral groups, let us suppose furthermore that

$$(RS)^2 \neq E . \qquad (16.10)$$

Then form a group H by adjoining to J a new involutory element
R_1 which transforms the elements of J according to the outer
automorphism just mentioned, that is

$$R_1R = RR_1, \quad R_1S = S^{-1}R_1 . \qquad (16.11)$$

From these relations it follows immediately that both of the products $R_1 R$ and $R_1 S$ are involutory, and, for agreement with (16.1) and (16.2), we might call them R_3 and R_2 respectively:

$$R_2 = R_1 S, \quad R_3 = R_1 R . \tag{16.12}$$

Hence the same group H can be generated by the three involutory elements R_1, R_2, R_3, thus giving rise to a trivalent Cayley graph, possibly of type ${}^3 z$.

Before considering examples—this will be done in the next sections—let us see what can be said about the parameters of this Cayley graph. (Recall their definitions given in Section 13.)

From (16.9) and (16.12) it follows that

$$(R_3 R_1)^2 = E, \quad (R_1 R_2)^2 \ne E , \tag{16.13}$$

and from (16.10) that also

$$(R_2 R_3)^2 \ne E ; \tag{16.14}$$

hence

$$p_3 > 2, \quad p_1 > 2, \quad p_2 = 2 . \tag{16.15}$$

The last equality means of course that our graph has girth 4. As to the two inequalities they might be replaced by the following more precise statement: if σ is the period of S and τ that of RS in J, then it follows from (16.2) and (16.4) that

$$p_3 = \max(\sigma, \tau), \quad p_1 = \min(\sigma, \tau) , \tag{16.16}$$

provided that we permute the generators whenever necessary to satisfy (13.2).

As to the parameter q, it follows from its definition as the period of the product $R_1R_2R_3$ that it must be even, say

$$q = 2\kappa .$$
(16.17)

Indeed, if q were odd, the relation

$$(R_1R_2R_3)^q = E$$
(16.18)

would give us a closed walk of length 3q, and thus with an odd number of edges in a Cayley graph which by construction is obviously bipartite—a contradiction!

Hence we have

$$(R_1R_2R_3)^q = (R_1R_2R_3R_1R_2R_3)^\kappa = (SRS^{-1}R)^\kappa ,$$
(16.19)

and we see that the parameter q is twice the period of the commutator

$$SRS^{-1}R = SRS^{-1}R^{-1} .$$
(16.20)

REMARK: Our procedure might also be explained in graph-theoretic terms this way. In the Cayley graph of J generated by R and S a new vertex is inserted in the middle of each S-edge, which thereby becomes an R_1R_2-path. Then the endpoints of each RR_1-path are joined by an R_3 edge; and finally the R-edges are removed. (See Fig. 16.1.)

On the other hand a reader familiar with the "duplication principle" (used by Frucht (1955) as a tool for obtaining graphs of type ^3S) will already have recognized that the procedure proposed here is essentially nothing else than the

Fig. 16.1   Construction of the Cayley graph of H from
            that of J.

case k = 2 of that principle, with the only difference that
now we are mainly interested in graphs of type ^{3}z.   The
occasional failure of that "principle"—see (Frucht, 1955, p.
413)—is not to be feared here because of the assumed existence
of the required automorphism of J taking R and S respectively
into their inverses.

17   THE CAYLEY GRAPHS OF THE GROUPS B(m,k)

Let $J(m,k)$ be the group defined by the following relations:

$$R^2 = S^m = E, \quad (RS)^2 = S^{k+1} , \tag{17.1}$$

In order to satisfy (16.9) and (16.10) only cases will be considered where

$$m > 2 \tag{17.2}$$

and

$$k < m - 1 ; \tag{17.3}$$

to avoid the uninteresting case of abelian groups let us suppose furthermore that

$$k > 1 . \tag{17.4}$$

Since the last relation in (17.1) is equivalent to

$$R^{-1}SR = S^k , \tag{17.5}$$

the consistency of the defining relations (17.1) requires that

$$k^2 \equiv 1 \pmod{m} . \tag{17.6}$$

Before using this group $J(m,k)$ as "group J" (in the procedure described in the foregoing section) the following remarks should prove useful.

(i)   The group $J(m,k)$ is of order $2m$; indeed it is nothing else than the special case $s = 2$ of the group of order ms defined by the relations (7.1) and (7.2).

(ii) Because of the conditions (17.3) and (17.4) we confine ourselves to the cases where

2 ≤ k ≤ m - 2;   (17.7)

then the groups J(m,k) always occur in pairs. Indeed, if an integer k satisfies both the conditions (17.6) and (17.7) so also does m-k. Although not isomorphic, both groups of such a pair have the same Cayley graph, namely the generalized Petersen graph called G(m,k) by Frucht et al. (1971) and P(m,k) by Biggs (1974). More interesting for us is the fact that the Cayley graphs with 4m vertices resulting from our procedure are companion graphs, as can easily be seen from Figs. 17.1 and 17.3 below. Therefore from any pair we shall always choose the group J(m,k) with the smaller value of k. In other words, the condition (17.7) will be replaced by the following one:

$$2 \le k < \frac{m}{2}.$$ (17.8)

(It is easily seen that the equality $k = \frac{m}{2}$ cannot hold here.)

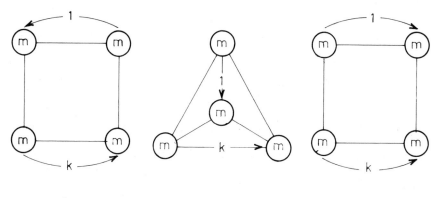

Fig. 17.1   Fig. 17.2   Fig. 17.3

Cayley graph, first companion and second companion of the group B(m,k) obtained from J(m,k).

It can be seen from Table 17.1 that for $m \leq 30$ there exist 11 groups $J(m,k)$ satisfying the condition (17.8). It should also be pointed out that in six cases, namely Nos. 1,2,4,5,9, 10, the following relation holds:

$$m = 2k + 2; \tag{17.9}$$

in the five remaining cases we have

$$m > 2k + 2 . \tag{17.10}$$

Using now $J(m,k)$ as starting point for the procedure described in the foregoing section we'll obtain a group of order $4m$ with three involutory generators subject to the following defining relations:

$$(R_1R_2)^m = (R_1R_3)^2 = E \tag{17.11}$$

and

$$(R_3R_2)^2 = (R_1R_2)^{k+1} . \tag{17.12}$$

This latter relation is also equivalent to

$$R_3R_2R_1R_3 = (R_1R_2)^k . \tag{17.13}$$

This group will be called $B(m,k)$ since the same notation has been used before by Frucht et al. (1971) in general for the "spokes-preserving" subgroup of the automorphism group of the generalized Petersen graph just mentioned. (From Frucht et al. (1971, Theorem 1b) it can be seen that this subgroup becomes our group with defining relations (17.11) and (17.12) when (17.6) is fulfilled.)

Before describing the Cayley graph of $B(m,k)$ let us determine its parameters. Because of (16.16) and (17.1) we have obviously

Table 17.1

Groups B(m,k) Satisfying (17.6) and (17.8), and their 0-Symmetric Cayley Graphs

No.	Order	m	k	Parameters	0-Symmetric Companion Graphs*	Group is Direct Product
1	32	8	3	8,4,2,8	none	—
2	48	12	5	12,4,2,6	1st	$D_4 \times D_3$
3	60	15	4	15,6,2,10	1st and 2nd	$D_5 \times D_3$
4	64	16	7	16,4,2,16	none	—
5	80	20	9	20,4,2,10	1st	$D_5 \times D_4$
6	84	21	8	21,14,2,6	1st and 2nd	$D_7 \times D_3$
7	96	24	5	24,8,2,12	1st and 2nd	—
8	96	24	7	24,6,2,8	1st and 2nd	$D_8 \times D_3$
9	96	24	11	24,4,2,24	none	—
10	112	28	13	28,4,2,14	1st	$D_7 \times D_4$
11	120	30	11	30,10,2,6	1st and 2nd	$D_6 \times D_5 \cong D_{10} \times D_3$

*not counted if isomorphic to the original Cayley graph

$$p_3 = m, \quad p_1 = \frac{2m}{(m,k+1)}, \quad p_2 = 2 . \tag{17.14}$$

When relation (17.9) holds the parameters are

$$p_3 = m, \quad p_1 = 4, \quad p_2 = 2. \tag{17.15}$$

As to the parameter q, it follows from (16.20) and (17.5) that q is twice the period of S^{k-1}; hence

$$q = \frac{2m}{(m,k-1)} . \tag{17.16}$$

It is easy to find a hamiltonian circuit in the Cayley graph of B(m,k) with respect to the generators R_1, R_2, R_3; indeed it follows from (17.11) that

$$(R_3 R_1 R_3 R_2)^m = E , \tag{17.17}$$

and this relation yields the LCF representation

$$[3, 4k+1, -(4k+1), -3]^m \tag{17.18}$$

with chord length number $N_c = 2$. (The "Rule 1" for LCF codes—mentioned in Section 3—is not violated since it follows from (17.8) that $4k+1 < 2m$.)

As can be seen from Table 17.1, the smallest instance is the graph with LCF code $[3, 13, -13, -3]^8$ for m = 8, k = 3; the parameters are 8,4,2,8. That this graph is ^3Z can be shown by the following argument: any vertex belongs to three different octagons, namely to those corresponding to the relations

$$(R_2 R_3)^4 = (R_2 R_3 R_2 R_1)^2 = (R_3 R_2 R_1 R_2)^2 = E ; \tag{17.19}$$

but of the three edges incident with that vertex only one is
common to the three octagons, one belongs to two of them, and
the third to only one octagon. Hence there are at each vertex three different kinds of incident edges and the stabilizer
of any vertex can only be trivial.

A similar reasoning holds for the other graphs satisfying
the relation (17.9), showing that they are 0-symmetric. For
the remaining graphs (Nos. 3,6,7,8,11 in Table 17.1) it has
been checked individually that they too are ^3Z.

Being of girth 4, the Cayley graph of B(m,k) has two
companion graphs (as described in Section 14) which are shown
in Frucht notation in Figs. 17.2 and 17.3 together with the
original Cayley graph in Fig. 17.1. Let us begin with the
discussion of the second one which turns out to be easier.
Here we have to use

$$R = R_1R_3 = R_3R_1 ,$$ (17.20)

instead of R_3, as the third involutory generator. Because of

$$RR_1RR_2 = R_1R_2 = R_3R_1R_3R_2$$ (17.21)

we can obtain—analogously to (17.17)—a hamiltonian circuit
from the relation

$$(RR_1RR_2)^m = E ,$$ (17.22)

giving us the LCF representation

$$[3,-(4k-1),4k-1,-3]^m .$$ (17.23)

Instead of (17.14) we now have the following values for the
parameters:

$$p_3 = m, \quad p_1 = \frac{2m}{(m,k-1)}, \quad p_2 = 2, \quad q = \frac{2m}{(m,k+1)} \ . \qquad (17.24)$$

These "second companion graphs" are also bipartite as can be seen from (17.23); they are however not necessarily 0-symmetric. Indeed, whenever the relation (17.9) holds, the second companion graph is not ^3Z, but ^3T. So there are only five 0-symmetric graphs with LCF code (17.23). The smallest one, with 60 vertices, corresponds to the group B(15,4), having parameters 15,10,2,6, and the code $[3,-15,15,-3]^{15}$.

Let us consider the "first companion graph" of the Cayley graph B(m,k), where R_2, R_3, and R—again defined by (17.20)— are used as the three involutory generators. The parameters are now

$$p_3 = \frac{2m}{(m,k-1)}, \quad p_1 = \frac{2m}{(m,k+1)}, \quad p_2 = 2, \quad q = m \ , \qquad (17.25)$$

where for the groups No. 6 and No. 11 the values of p_3 and p_1 should be interchanged in order to have $p_3 \geq p_1$, according to (13.2).

In the discussion of these graphs it will be convenient to distinguish three cases.

Case I: The relation (17.9) is satisfied, and we also have

$$m \equiv 0 \pmod 8 \ . \qquad (17.26)$$

(In Table 17.1 this happens for Nos. 1,4, and 9). Then it is easily checked that no new graphs arise, because in this case the first companion graph turns out to be isomorphic to the original Cayley graph with LCF code (17.18).

Case II: Also now the relation (17.9) holds, but with

$$m \equiv 4 \pmod 8 . \tag{17.27}$$

(Nos. 2,5,10 in Table 17.1).   In this case the parameters (17.25) become

$$p_3 = \frac{m}{2}, \ p_1 = 4, \ p_2 = 2; \ q = m ; \tag{17.28}$$

that is, we have the relations

$$(RR_2)^{m/2} = (R_2R_3)^4 = (R_3R)^2 = (RR_2R_3)^m = E . \tag{17.29}$$

It follows that

$$R_2 = (R_3R_2)^3 R_3 \tag{17.30}$$

and

$$\{(R_3R_2)^3 R_3R\}^{m/2} = E . \tag{17.31}$$

The last relation yields a hamiltonian circuit with chord length number $N_c = 4$ and the LCF representation

$$[7,-3,11,9;-]^{m/2} \tag{17.32}$$

(for m = 12,20,28,...) from which it is seen that in Case II the graph is bipartite.   That it is 0-symmetric can be shown by an argument based on the number of octagons passing through the three edges incident with any vertex, similar to that used above for B(8,3).

Case III: The relation (17.9) is not satisfied; that is, the inequality (17.10) holds.   (In Table 17.1 this occurs for Nos. 3,6,7,8 and 11.)   Here we have found no uniform way to proceed, except by referring to the Frucht notation for these graphs as shown in Fig. 17.2. Nevertheless, in each case there is a relation of the form

$$(R_a R_b (R_a R_c)^s)^p = E, \quad \{a,b,c\} = \{0,2,3\} , \qquad (17.33)$$

(where we write R_0 for R) that does produce the desired hamiltonian circuit. In Table 17.2 we list for each group the word in relation (17.33) and the resulting LCF code. The five graphs are all 0-symmetric. We note that the first two of these are not bipartite.

Table 17.2   LCF Codes for some "First Companion" Graphs

No. in Table 17.1	m	k	Word	LCF Code
3	15	4	$R_2 R (R_2 R_3)^2$	$[30,9,7;-]^{10}$
6	21	8	$R_2 R_3 (R_2 R)^2$	$[42,9,7;-]^{14}$
7	24	5	$R_3 R (R_3 R_2)^3$	$[-41,-3,11,9;-]^{12}$
8	24	7	$R_3 R (R_3 R_2)^5$	$[11,-3,19,17,-9,-11;-]^8$
11	30	11	$R R_3 (R R_2)^5$	$[11,-3,19,17,-9,-11;-]^{10}$

Let us consider the second companion graph of B(15,4), with parameters 10,6,2,15. The group may be represented in terms of permutations by using as generators

$$\left. \begin{aligned} R &= (1\ 5)(2\ 4) , \\ R_2 &= (1\ 4)(2\ 3)(6\ 7) , \\ R_3 &= (7\ 8) . \end{aligned} \right\} \qquad (17.34)$$

This representation clearly corresponds to the result of the juxtaposition procedure, explained in Section 13, applied to a dihedral group D_5 with generators (1 5)(2 4) and (1 4)(2 3) and a group $D_3 \cong S_3$ with generators (6 7) and (7.8). The group B(15,4) is in fact isomorphic to the direct product $D_5 \times D_3$, and a similar isomorphism holds for several of these groups. (See the last column of Table 17.1.) Indeed, we shall prove the following.

THEOREM 17.1: If m is the product of two coprime integers u and v, both greater than 2, then there exists an integer k satisfying the conditions (17.6) and (17.8) such that

$$B(m,k) \cong D_u \times D_v . \tag{17.35}$$

PROOF: If u and v satisfy the conditions stated in the theorem, namely

$$(u,v) = 1, \; u > 2, \; v > 2 , \tag{17.36}$$

then the direct product $D_u \times D_v$ of order 4uv can be generated by three involutory elements R_1, R_2, R_3 as follows:

$$R_1 = U_1V_1, \; R_2 = U_2V_2, \; R_3 = U_1 , \tag{17.37}$$

where U_1, U_2, V_1, V_2 are supposed to satisfy the relations

$$U_i^2 = V_j^2 = E, \; U_iV_j = V_jU_i \quad (i, j = 1,2) \tag{17.38}$$

and

$$(U_1U_2)^u = (V_1V_2)^v = E . \tag{17.39}$$

It is easily checked that then the defining relations (17.11) and (17.12) of the group B(m,k) of the same order 4m are satisfied, with

$$m = uv , \tag{17.40}$$

if k is a solution of the simultaneous congruences

$$k \equiv 1 \pmod{u} \tag{17.41}$$

$$k \equiv -1 \pmod{v} . \tag{17.42}$$

But it follows from the so-called Chinese remainder theorem that these congruences have just one solution satisfying $2 \leq k < m$; if this solution does not satisfy (17.8), replace it by m-k. (This is possible because it means only that we interchange u and v; recall that neither u nor v can be 2.) In each case, by squaring the congruences (17.41) and (17.42) we see that the value of k thus found satisfies also the condition (17.6). #

REMARK: Surprisingly Fig. 17.2 represents also the Cayley graph of the group Z(m,2,k) defined by the generating relations (8.2). This might seem to be a contradiction to Theorem 4.2, according to which a 0-symmetric graph cannot be the Cayley graph of two non-isomorphic groups at the same time. There is however no such contradiction, since the definition (8.2) of Z(m,2,k) requires that k satisfy the congruence

$$k^2 \equiv -1 \pmod{m} ; \tag{17.43}$$

but now in B(m,k) we have

$$k^2 \equiv 1 \pmod{m} . \tag{17.44}$$

In this section some graphs of type ^3z and girth 4 which are related to those of the foregoing section will be considered.

I. It has been shown above (see Theorem 17.1 and Table 17.1) that any direct product of two dihedral groups with coprime degrees greater than 2 is a group B(m,k) and as such may have one or more 0-symmetric Cayley graphs of girth 4. There are however also direct products of dihedral groups whose degrees are not coprime and which nevertheless can be generated by three involutory elements, giving rise to 0-symmetric Cayley graphs of girth 4. Within the range of our study the following cases have been found.

(1) The group p[4] r is generated by two elements T and S of orders p and r respectively, which satisfy $(TS)^2 = (ST)^2$. Coxeter (1974, pp. 106-107) implies, without stating it explicitly, that p[4]r is isomorphic to the direct product p[r]p × C_r when r is odd. The case of interest to us is p = 2, for which the generators of 2[4]r satisfy

$$R^2 = S^r = E, \quad (RS)^2 = (SR)^2 . \tag{18.1}$$

(Here we have used R in place of T for the now involutory generator.) Furthermore, when r is odd we have the group 2[4]r, of order $2r^2$, isomorphic to the direct product $D_r \times C_r$. Since (18.1) shows that this group admits an automorphism taking each generator into its inverse, we may use it as "group J" in the procedure of Section 16. The group H,

which results from adjoining the involutory element R_1 that
effects this automorphism (see (16.12)), is defined by

$$(R_1R_3)^2 = (R_1R_2)^r = (R_1(R_2R_3)^2)^2 = E . \tag{18.2}$$

It is easy to show that H is the direct product $D_r \times D_r$;
involutory generators for the two factors might be

$$R_3, (R_3R_2)^r R_2 \text{ and } R_1R_3, (R_3R_2)^r . \tag{18.3}$$

These observations are amply illustrated by the permutation
representation (for $r = 3$)

$$R_1 = (1\ 2)(4\ 6), \ R_2 = (1\ 3)(4\ 5), \ R_3 = (1\ 2) . \tag{18.4}$$

From (18.2), one can easily see that the parameters of
our group (rearranged to conform with (13.2)) are $2r$, r, 2,
$2r$. In the case $r = 3$, the parameters are the same as those
of the group $G^{3,6,6}$ (No. 3b of Section 15), of order 108.
This should not be astonishing since $D_3 \times D_3 \cong S_3 \times S_3$ is a
factor group of $G^{3,6,6}$ (Coxeter 1977, p. 266).

In the Cayley graph of $D_r \times D_r$, the relation

$$((R_1R_2)^{r-1}R_1R_3)^{2r} = E , \tag{18.5}$$

provides a hamiltonian circuit from which we find the LCF
code

$$[5,-3,-17;-]^6 \tag{18.6}$$

with $N_c = 3$ for $r = 3$, and

$$[9,-3,-25,-27,-49;-]^{10} \tag{18.7}$$

with $N_c = 5$ for $r = 5$. These are the only two graphs of this
family that meet our size restrictions.

Both graphs are easily seen to be ^3Z. At each vertex, the R_2-edge lies on no square, and the R_1- and R_3-edges lie on different numbers of 2r-gons.

(2)   Because of the well-known fact that $D_r \times C_2 \cong D_{2r}$ when r is odd, we may apply Theorem 13.2 to the direct products above, to obtain $D_r \times D_r \times C_2 \cong D_{2r} \times D_r$ of order $8r^2$, using as generators

$$R_1 = R_1', \ R_2 = R_2'P, \ R_3 = R_3' . \tag{18.8}$$

(The R_i' are the generators of $D_r \times D_r$ satisfying (18.2), and P generates C_2.)   The parameters are 2r, 2r, 2, 2r.   For the case r = 3—the only one with not more than 120 elements—the graph is 0-symmetric, and the relation

$$((R_1R_2)^5R_1R_3)^6 = E \tag{18.9}$$

yields a hamiltonian circuit with $N_c = 6$ and the LCF code

$$[11,-3,-29,-31,15,13;-]^6 . \tag{18.10}$$

Another LCF code, with the smaller chord length number $N_c = 4$, is available, namely

$$[11,-3,15,13,-13,-15;-]^6 , \tag{18.11}$$

coming from the relation

$$((R_3R_2)^5R_3R_1)^6 = E . \tag{18.12}$$

Neither this graph, nor either of the two preceeding, can have a new 0-symmetric companion graph.

(3)   The direct product $D_9 \times D_3$ of order 108 can be generated
by the permutations

$$R_1 = (1\ 9)(2\ 8)(3\ 7)(4\ 6)(10\ 11),$$

$$R_2 = (2\ 9)(3\ 8)(4\ 7)(5\ 6)(10\ 12),$$ (18.13)

$$R_3 = (10\ 11),$$

The parameters are 9, 6, 2, 18, and from the relation

$$((R_1R_2)^8R_1R_3)^6 = E$$ (18.14)

we can obtain a hamiltonian circuit with the LCF code

$$[17,-3,-41,-43,27,25,-13,-15,-53;-]^6 .$$ (18.15)

Not only this graph is ^3Z, but also its two companion
graphs, having respectively the parameters 18, 6, 2, 9 and
18, 9, 2, 6.   In the first, the relation

$$((R_2R)^8R_2R_3)^6 = E$$ (18.16)

gives a hamiltonian circuit with LCF code

$$[54,33,31,-7,-9,25,23,-15,-17;-]^6 .$$ (18.17)

In the second we were unable to find a "nice" hamiltonian
circuit; however, it is easy to draw this graph by applying
the definitions given in Section 14.

II.   The direct products of dihedral groups for which we obtained Cayley graphs of type ^3Z were all of the form $D_u \times D_v$
with at least one of the numbers u, v odd.   One might be
tempted to apply the juxtaposition procedure also to direct
products of the form $D_u \times D_v$ with both u and v even; it turns

out, however, that in this case only a subgroup of index 2
will be obtained. Notwithstanding, such cases can be sources
of new 0-symmetric graphs as can be seen from the following
instance.

(4) A group of order 64 can be obtained by using the generators

$$
\left.
\begin{aligned}
R_1 &= (1\ 8)(2\ 7)(3\ 6)(4\ 5)\ , \\
R_2 &= (2\ 8)(3\ 7)(4\ 6)(10\ 12)\ , \\
R_3 &= (9\ 12)(10\ 11)\ .
\end{aligned}
\right\}
\qquad (18.18)
$$

Since all the generators are even permutations, it is rather
obvious that this group cannot be the whole direct product
$D_8 \times D_4$. The parameters are 8, 4, 2, 8, i.e., equal to those
of the Cayley graph of the group B(8, 3) of order 32 considered in Section 17. In fact B(8, 3) is a homomorphic image
of the group we are considering now. The relation

$$
((R_3R_2)^3 R_3R_1)^8 = E \qquad (18.19)
$$

gives a hamiltonian circuit with chord length number $N_c = 4$
and LCF code

$$
[7,-3,11,9;-]^8\ , \qquad (18.20)
$$

and it is easily checked (e.g., by counting the number of
octagons containing a given edge) that this graph is ^3z.

The reader should not be confused by the resemblance
between (18.20) and the LCF code (17.32) obtained above for
the first companion graph of the Cayley graph of certain
groups B(m,k). The difference resides in the fact that in

(17.32) the exponent $\frac{m}{2}$ was restricted to the values 6, 10, 14,..., while now in (18.20) the exponent is 8; and it might also be 12, 16, 20,... . Of this rather obvious generalization of (18.18) we mention the following case, the only one falling into the range of our study.

(5)   A group of order 96, with parameters 12, 4, 2, 12 is obtained by using as generators

$$
\left.
\begin{aligned}
R_1 &= (1\ 12)(2\ 11)(3\ 10)(4\ 9)(5\ 8)(6\ 7) \ , \\
R_2 &= (2\ 12)(3\ 11)(4\ 10)(5\ 9)(6\ 8)(14\ 16) \ , \\
R_3 &= (13\ 16)(14\ 15) \ .
\end{aligned}
\right\}
\qquad (18.21)
$$

The relation

$$
((R_3R_2)^3R_3R_1)^{12} = E \qquad\qquad (18.22)
$$

yields the same LCF code as (18.20), but with exponent 12. The graph has been verified to be 0-symmetric. Some other instances of direct products (but not related to dihedral groups) will be considered in the next section.

19   MORE 0-SYMMETRIC GRAPHS OF GIRTH 4
WITH 96 OR 120 VERTICES

I. We now return to the groups $F^{a,b,c}$ with the defining relations

$$R^2 = RS^a RS^b RS^c = E ,$$
(19.1)

already considered (see Section 12) as a source of Cayley graphs of type $^1 Z$, and ask the following question: "Are there members of the same family $F^{a,b,c}$ that might be used as 'group J' in the procedure described in Section 16, thus leading to groups with twice the order of J and with Cayley graphs of type $^3 Z$?"

Recalling what was said above after (16.9), we are of course only interested in groups admitting an outer automorphism which leaves R fixed and changes S into S^{-1}. It turns out that this condition is rather restrictive, since the following theorem is easily proved.

THEOREM 19.1: The existence of an automorphism of $F^{a,b,c}$ leaving R fixed and taking S into S^{-1} implies that S^d is a central element, where d is the greatest common divisor of $|a-b|$, $|b-c|$ and $|c-a|$.

PROOF: The existence of an automorphism as described in the theorem means that the relation

$$RS^{-a} RS^{-b} RS^{-c} = E$$
(19.2)

is also valid in $F^{a,b,c}$; hence

$$S^{-b} RS^{-c} = RS^a R .$$
(19.3)

107

But it follows from (19.1) that

$$RS^a R = S^{-c}RS^{-b} \; ; \qquad\qquad (19.4)$$

whence we have the relation

$$S^{-b}RS^{-c} = S^{-c}RS^{-b} \; , \qquad\qquad (19.5)$$

which is equivalent to

$$RS^{b-c} = S^{b-c}R \; . \qquad\qquad (19.6)$$

Thus R commutes with S^{b-c}, and also with S^{c-a} and S^{a-b} (as is shown in an analogous fashion), and so also with S^d where

$$d = (|b-c|, \; |c-a|, \; |a-b|) \; . \qquad\qquad (19.7)$$

In other words, S^d is a central element as was to be shown. #

Within the range of our study we found only two such groups $F^{a,b,c}$ that can be used as "group J" (in the sense of Section 16), and since both groups are of the same order 48 (but not isomorphic!), the procedure described in Section 16 will allow us to find two groups of order 96 with several Cayley graphs that turn out to be of type ^3Z (and are, of course, of girth 4). Surprisingly, both groups of order 96 turn out to be isomorphic to the group H of collineations and correlations of the Möbius-Kantor configuration 8_3, studied by Coxeter (1977), as we shall show.

The group H of collineations and correlations of 8_3 is generated by two elements B, C, with the presentation (Coxeter 1977, p. 297)

$$(BC)^2 = (B^2C^3)^2 = (B^3C^2)^2 = (B^{-1}C^2)^2 = E \; . \qquad\qquad (19.8)$$

It follows from these relations that

$$A^3 = B^8 = C^{12} = (AB)^2 = (CA)^2 = (BC)^2 = (ABC)^2 = E \quad (19.9)$$

(where $A = B^2C^2$). Thus we see that our group is a homomorphic image of the infinite group $G^{3,8,12}$ (see 14.8). With a slight change in notation from Section 14, we define three elements

$$R_1 = AB, \ R_2 = B^2CB^{-1}, \ R_3 = BC \ . \qquad (19.10)$$

The relations (19.8) show that these are involutory and that they generate H, since

$$A = R_2R_3, \ B = R_3R_2R_1, \ C = R_1R_2 \ . \qquad (19.11)$$

Moreover, in conformity with (14.1), we have

$$R_1R_3 = R_3R_1 = R = CA \ . \qquad (19.12)$$

From Coxeter (1977, pp. 294-295) we have for A, B and C these permutations of 16 symbols (eight numerals and eight letters, representing the points and lines of the self-dual configuration 8_3):

$$
\left.
\begin{array}{l}
A = (1 \ 7 \ 4)(2 \ 5 \ 8)(a \ b \ f)(c \ h \ g) , \\
B = (1 \ d \ 4 \ g \ 8 \ e \ 5 \ b)(2 \ h \ 3 \ c \ 7 \ a \ 6 \ f) , \\
C = (1 \ h \ 8 \ a)(2 \ c \ 6 \ g \ 4 \ d \ 7 \ f \ 3 \ b \ 5 \ e) .
\end{array}
\right\} \quad (19.13)
$$

Consequently, the involutory generators have this representation:

$$
\left.
\begin{array}{l}
R_1 = (1 \ a)(2 \ b)(3 \ c)(4 \ d)(5 \ e)(6 \ f)(7 \ g)(8 \ h) , \\
R_2 = (2 \ 5)(3 \ 6)(4 \ 7)(a \ h)(b \ c)(f \ g) , \\
R_3 = (1 \ 7)(2 \ 8)(3 \ 6)(a \ g)(b \ h)(c \ f) , \\
R \ = (1 \ g)(2 \ h)(3 \ f)(4 \ d)(5 \ e)(6 \ c)(7 \ a)(8 \ b) .
\end{array}
\right\} \quad (19.14)
$$

It is known (Coxeter 1977, p. 295) that A and C generate a subgroup isomorphic to $3[6]2 \cong F^{4,1,-2}$. More precisely, the elements

$$R = R_1R_3 = CA, \quad S = R_1R_2 = C \qquad (19.15)$$

satisfy the defining relations for $F^{4,1,-2}$. Hence we see that H is indeed isomorphic to the extension of $F^{4,1,-2}$ by the procedure of Section 16.

Similarly it is known (Coxeter 1977, p. 295) that A and B generate a subgroup of H isomorphic to the unimodular group mod 3, $GL(2,3) \cong F^{1,1,-3}$. To be exact, the elements

$$\bar{R} = RR_3 = AB, \quad \bar{S} = (R_3R_2RR_3)^3 = B^3 \qquad (19.16)$$

satisfy the defining relations of $F^{1,1,-3}$. Since $\bar{R}$ and $\bar{S}$ are products of even numbers of the generators R, R_2, R_3, we see that H is also isomorphic to the extension of $F^{1,1,-3}$ by the procedure of Section 16. We see furthermore that the first companion of the Cayley graph of the extension of $F^{1,1,-3}$ as generated by $\bar{R}$ and $\bar{S}^3 = B$ is isomorphic to the Cayley graph of the extension of $F^{4,1,-2}$ as generated by R and S.

Now let us study the graphs of the group H. We start from $F^{4,1,-2}$ with defining relations

$$R^2 = RS^4RSRS^{-2} = E \qquad (19.17)$$

in which S^3 is central. (See Campbell et al. 1977, p. 434.) From equations (19.9, 19.11, 19.12) we easily determine that

$$(R_1R_2)^{12} = (R_2R_3)^3 = (R_3R_1)^2 = (R_1R_2R_3)^8 = E , \qquad (19.18)$$

showing that this Cayley graph of H has parameters 12,3,2,8.
The relation

$$(R_3R_2R_1R_3R_1R_2R_3R_1)^{12} = E \qquad\qquad (19.19)$$

yields a hamiltonian circuit with chord-length number $N_c = 3$
and the LCF representation

$$[7,-3,3,-31;-]^{12} . \qquad\qquad (19.20)$$

In the first companion graph (with parameters 8,3,2,12)
the relation

$$(R_3R_2R(R_2R_3)^2R_2RR_2R_3R)^8 = E \qquad\qquad (19.21)$$

gives a hamiltonian circuit with $N_c = 6$ and LCF code

$$[11,-3,7,5,39,37;-]^8 . \qquad\qquad (19.22)$$

In the second companion graph, with parameters 12,8,2,3 and
hence not bipartite, we obtain the LCF representation

$$[-41,-3,34,34;-]^{12} \qquad\qquad (19.23)$$

from the relation

$$((RR_2)^3RR_1)^{12} = E . \qquad\qquad (19.24)$$

All three graphs are 0-symmetric.
   We now turn to the group $F^{1,1,-3}$ with defining relations

$$\bar{R}^2 = \bar{R}\,\bar{S}\,\bar{R}\,\bar{S}\,\bar{R}\,\bar{S}^{-3} = E \qquad\qquad (19.25)$$

(Overbars are used to distinguish the generators of this group
and H as its extension.) Here $\bar{S}^4$ is central and also involutory. (See Campbell et al. 1977, pp. 432-433.) By (19.16),

the generators of H satisfy

$$\bar{R}_1\bar{R}_3 = \bar{R} = AB, \quad \bar{R}_1\bar{R}_2 = \bar{S} = B^3 \tag{19.26}$$

from which we find that the parameters of the Cayley graph
are 8,6,2,12. The relation

$$(\bar{R}_3\bar{R}_2\bar{R}_1(\bar{R}_2\bar{R}_3)^2\bar{R}_2\bar{R}_1\bar{R}_2\bar{R}_3\bar{R}_1)^8 = E \tag{19.27}$$

(analogous to (19.21)) furnishes a hamiltonian circuit on
which the chord-length number is again $N_c = 6$ and the LCF
code is

$$[11,-3,-41,-43,39,37;-]^8 . \tag{19.28}$$

In the first companion graph, the parameters are 12,6,2,8;
the relation

$$(\bar{R}_3\bar{R}_2\bar{R} \bar{R}_3\bar{R} \bar{R}_2 \bar{R}_3\bar{R})^{12} = E \tag{19.29}$$

(analogous to (19.19)) gives a hamiltonian circuit and thence
the LCF representation

$$[-41,-3,3,17;-]^{12} \tag{19.30}$$

with $N_c = 3$. In the second companion graph, with parameters
12,8,2,6, the LCF code

$$[-41,-3,-14,-14;-]^{12} \tag{19.31}$$

comes from the relation

$$(\bar{R}_1\bar{R}(\bar{R}_1\bar{R}_2)^3)^{12} = E \tag{19.32}$$

and shows the graph non-bipartite. In (19.31) we might shift
the starting point of the hamiltonian circuit to obtain the
more pleasant form

$$[14,14,3,41;-]^{12} \, .\tag{19.33}$$

These last three graphs are also 0-symmetric; thus we have obtained a total of six nonisomorphic graphs, each of which represents the group of collineations and correlations of 8_3.

II. An instance where a 0-symmetric Cayley graph of girth 4 results from applying Theorem 13.2 and/or Corollary 13.2.1 <u>twice</u> is the following. Using the three involutory generators

$$R_1 = (1 \ 4), \ R_2 = (2 \ 4)(5 \ 6), \ R_3 = (1 \ 4)(2 \ 3)(7 \ 8)\tag{19.34}$$

for the direct product $S_4 \times C_2 \times C_2$ of order 96 we obtain a Cayley graph with parameters 6,4,2,6, for which the relation

$$((R_2R_3)^3 R_2 R_1 (R_2R_3)^2 R_1 R_3 R_2 R_1)^6 = E\tag{19.35}$$

yields a hamiltonian circuit with the (not antipalindromic) LCF code

$$[7,41,39,-13,-15,15,13,-7,-9,-39,-41,3,-15,15,-3,9]^6 \, .$$
$$\tag{19.36}$$

Of the two companion graphs, one is isomorphic to the graph described, while the other (with parameters 6,6,2,4) is not 0-symmetric.

III. A Cayley graph of girth 4 with 120 vertices can be obtained by a combination of the juxtaposition procedure with the method ("duplication principle") explained in Section 16.

The direct product $A_4 \times C_5$ of order 60 can be generated by the following permutations of nine symbols:

$$R = (1 \ 2)(3 \ 4), \ S = (1 \ 2 \ 3)(5 \ 6 \ 7 \ 8 \ 9) \, .\tag{19.37}$$

Since the replacement of R and S by their inverses yields an automorphism, we can use (19.37) as generators of a group J in the sense of Section 16. Then by adjoining to J an involutory element R_1, via (16.11) and (16.12), we obtain a group H of order 120 with three involutory generators. The corresponding Cayley graph with parameters 15,15,2,4 is not 0-symmetric, but its first companion graph (in the sense of Section 14) with parameters 15,4,2,15 is ^3Z (as is easily checked). The relation

$$((RR_2)^3RR_3)^{15} = E \qquad\qquad (19.38)$$

yields a hamiltonian circuit with the chord length number $N_c = 3$ and the LCF code

$$[7,-3,10,10;-]^{15} , \qquad\qquad (19.39)$$

differing only in the exponent from (15.13).

IV. The symmetric group S_5 of order 120 can be generated by three involutory generators, e.g. by

$$R_1 = (1\ 2),\ R_2 = (2\ 3)(4\ 5),\ R_3 = (3\ 4) , \qquad\qquad (19.40)$$

giving rise to a Cayley graph with parameters 6,4,2,5. For the sake of brevity let

$$V = (R_3R_2)^2R_1R_2(R_3R_2)^3R_1R_2 ; \qquad\qquad (19.41)$$

then the relation

$$(R_3R_1R_3R_2R_3R_2R_1R_2VR_1R_2R_3R_1V)^3 = E \qquad\qquad (19.42)$$

gives us a hamiltonian circuit with the (not antipalindromic)
LCF representation

$$[60,37,35,-7,-9,-15,-17,3,-23,15,-3,23,21,9,7,56,56,$$
$$-29,-31,60,60,-7,-9,3,-15,23,-3,31,29,17,15,-56,-56,$$
$$-21,-23,9,7,-35,-37,60]^3 \ . \tag{19.43}$$

Both this graph and its two companion graphs are ^3z.

In Section 14 we introduced the companion graphs of Cayley graphs of groups with three involutory generators, two of which commute. These companion graphs arise from a suitable change of generators preserving the property that the Cayley graph has girth 4. It will be shown in this section that sometimes a different change of generators will allow us to derive from such a graph another kind of companion graph which is of higher girth and possibly 0-symmetric. The procedure is as follows.

As in Section 14, let us start with a group H having three involutory generators R_1, R_2, R_3, and suppose that $p_2 = 2$, i.e.,

$$R_1 R_3 = R_3 R_1 . \tag{20.1}$$

Since also the element

$$T_1 = R_2 R_1 R_2 = R_2^{-1} R_1 R_2 \tag{20.2}$$

is involutory, we can generate the same group H equally well by T_1, R_2, R_3, and consider the corresponding Cayley graph. It is easily checked that of the four parameters (introduced in Section 13) only p_2 is possibly affected by this change of generators. By definition, indeed, the new value of p_2 (if the condition (13.2) is not taken into account) will be equal to the period of

$$T_1 R_3 = R_2 R_1 R_2 R_3 , \tag{20.3}$$

and this period <u>can</u> be greater than 2. If this is the case
we obtain a new companion graph of girth g > 4. On the other
hand the girth of this new graph is at most 8, because of the
relation

$$(T_1 R_2 R_3 R_2)^2 = E ,$$ (20.4)

which is an immediate consequence of (20.1) and (20.2).

Another possibility for obtaining a companion graph is
that of generating the group H by U_1, R_2, R_3, where

$$U_1 = R_3^{-1} T_1 R_3 = R_3 R_2 R_1 R_2 R_3 .$$ (20.5)

If the new parameters p_2 and p_3 are both greater than 2 we
obtain a possibly 0-symmetric Cayley graph of girth g > 4
(and sometimes g > 8 as well be seen from the examples below).
By these two procedures the following 17 graphs of type ^3Z,
falling within the range of our study, were found.

I.   The group $G^{3,5,10} \cong A_5 \times C_2$ of order 120 has two 0-
symmetric Cayley graphs of girth 8. The first, with parameters 10, 6, 5, 3 (and hence not bipartite), comes from
changing the generators (15.34) to

$$T_1 = (1\ 2)(3\ 5),\ R_2 = (2\ 3)(4\ 5)(6\ 7),$$
$$R_3 = (1\ 2)(3\ 4)(6\ 7) .$$ (20.6)

The other, with parameters 10, 6, 5, 10, can be derived from
(15.38) by passing to the new generators

$$T_1 = (1\ 3)(2\ 5),\ R_2 = (2\ 3)(4\ 5)(6\ 7) ,$$
$$R_3 = (1\ 4)(2\ 3) .$$ (20.7)

II.   From (19.40) we obtain the new generators

$$T_1 = (1\ 3),\ R_2 = (2\ 3)(4\ 5),\ R_3 = (3\ 4) \tag{20.8}$$

for the symmetric group S_5 of order 120. The Cayley graph
has parameters 6, 4, 3, 5 (hence it is not bipartite!) and is
of girth 6 because of the relation

$$(T_1 R_3)^3 = E\ . \tag{20.9}$$

For the same group S_5 we can obtain two Cayley graphs of
girth 8 by applying the same procedure to the companion graphs
of (19.40). For the first we have the generators

$$T_1 = (1\ 3)(2\ 5),\ R_2 = (2\ 3)(4\ 5)\ ,$$
$$R_3 = (3\ 4) \tag{20.10}$$

and the parameters 6, 5, 4, 6; for the second we have

$$T_1 = (1\ 3),\ R_2 = (2\ 3)(4\ 5),\ R_3 = (1\ 2)(3\ 4) \tag{20.11}$$

and the parameters 6, 5, 4, 4.

III.   For the groups $B(m,k)$ discussed in Section 17 we found
(within the range of our study) twelve 0-symmetric companion
graphs of girth 8 or 10. These are summarized in Table 20.1
below.

TABLE 20.1

Companion Graphs of Groups B(m,k) of Higher Girth

No.	Group	Order	Generator[1]	Graph[2]	Parameters	Girth	LCF Code	Ref.[3]
1	B(15,4)	60	T_1,U_1	2	15,10,10,6	8	$[27,9;-]^{15}$	3
2	B(20,9)	80	T_1,U_1	2	20,10,10,4	8	$[-13,29;-]^{20}$	5
3	B(21,8)	84	T_1	-	21,14,14,6	8	$[-21,9;-]^{21}$	6
4	B(21,8)	84	U_1	-	21,14,14,6	8	$[15,-39;-]^{21}$	6
5	B(24,5)	96	T_1,U_1	-	24,8,4,12	8	$[27,45;-]^{24}$	7
6	B(24,5)	96	T_1	2	24,12,6,8	8	$[29,39,31,17,-39,-29;-]^{8}$	7
7	B(24,5)	96	U_1	2	12,8,6,8	10	$[35,13;-]^{24}$	7
8	B(24,7)	96	T_1,U_1	2	24,8,4,6	8	$[-21,45;-]^{24}$	8
9	B(24,11)	96	U_1	2	24,12,8,4	10	$[-13,37;-]^{24}$	9
10	B(28,13)	112	T_1	2	28,14,14,4	8	$[19,-35;-]^{28}$	10
11	B(28,13)	112	U_1	2	28,14,14,4	10	$[-13,45;-]^{28}$	10
12	B(30,11)	120	T_1,U_1	-	30,10,10,6	8	$[27,-51;-]^{30}$	11

(1) Shows whether T_1 (20.2) or U_1 (20.5) replaces R_1. (In five cases isomorphic graphs result from using T_1 or U_1.) (2) The number 2 means that the change of generators was made in the second companion graph. (Making it in the first companion graph does not change girth.)
(3) Number of the group in Table 17.1.

21 PRELIMINARIES ON CAYLEY GRAPHS OF GIRTH 6
FOR DIHEDRAL AND RELATED GROUPS

As in Section 16, let us consider a group H with involutory generators R_1, R_2, R_3, and the subgroup J formed by
those elements of H that can be written as products of an even
number of generators. To avoid Cayley graphs of girth 4,
however, we drop the condition (16.1) that R_1 and R_3 commute,
replacing it with the condition that the subgroup J be abelian.

In other words, we now write Y (not involutory) instead of
R for the product R_1R_3, i.e.,

$$R_1R_2 = S, \quad R_1R_3 = Y , \tag{21.1}$$

and we require that

$$SY = YS . \tag{21.2}$$

To avoid repetition of Section 16 and eliminate some non-0-
symmetric graphs, we also require that

$$p_i > 2, \quad i = 1,2,3 , \tag{21.3}$$

which means, in terms of S and Y, that

$$S^2 \neq E, \quad Y^2 \neq E, \quad (S^{-1}Y)^2 = (R_2R_3)^2 \neq E . \tag{21.4}$$

It is then obvious that

$$R_1^{-1}SR_1 = S^{-1}, \quad R_1^{-1}YR_1 = Y^{-1} . \tag{21.5}$$

Consequently the subgroup of H generated by S and R_1 is isomorphic to a dihedral group, as is that generated by Y and
R_1.

We already have required in (21.3) that there be no
squares in the graph corresponding to relations of the type

$$(R_\lambda R_\mu)^2 = E, \quad \lambda \neq \mu .$$
(21.6)

If any other four-letter word in the generators should equal
the identity, the Cayley graph of H would again have girth 4
and could not be 0-symmetric.   Indeed, we should then have a
relation of the form

$$R_\lambda R_\mu R_\lambda = R_\nu, \quad \{\lambda,\mu,\nu\} = \{1,2,3\}$$
(21.7)

and the interchange of generators R_μ and R_ν would yield a
graph automorphism leaving E fixed.

Eliminating these cases once and for all, we find that
the Cayley graph of H —being obviously bipartite— has girth
6, because of relations such as

$$(R_2 R_1 R_3)^2 = E ,$$
(21.8)

which follows directly from (21.2).   In fact, each vertex of
the graph belongs to at least the three hexagons corresponding
to the six relations

$$(R_\lambda R_\mu R_\nu)^2 = E, \quad \{\lambda,\mu,\nu\} = \{1,2,3\} .$$
(21.9)

(Each hexagon is described by two relations in two different
senses.)

In the case that each vertex belongs to only these three
hexagons, we may define a Petrie polygon as a circuit in which
no three consecutive edges lie on the same hexagon.   (See
Boreham et al. 1974, Boreham 1974, Coxeter 1950.)   From (21.9)
we see that if three consecutive edges on a path correspond

to three different generators, they do indeed belong to the same hexagon. Thus there are three Petrie polygons through each point of the graph, each corresponding to one of the relations

$$(R_1R_2)^{p_3} = E, \quad (R_2R_3)^{p_1} = E, \quad (R_3R_1)^{p_2} = E , \qquad (21.10)$$

having lengths $2p_3$, $2p_1$, $2p_2$, respectively.

Since a Petrie polygon has a topological definition, any automorphism of the Cayley graph of H must take a Petrie polygon into a Petrie polygon of the same length. We conclude, then, the following.

THEOREM 21.1: Let H be a group with three involutory generators satisfying (21.1), (21.2), (21.3), and let the Cayley graph of H have girth 6, each vertex lying on exactly three hexagons. Then any graph automorphism that leaves the vertex corresponding to E fixed is induced by a group automorphism that permutes the generators. #

From this theorem there follows easily this corollary.

COROLLARY 21.1.1: The Cayley graph of a group satisfying the hypotheses of Theorem 21.1 is 0-symmetric if the three parameters p_1, p_2, p_3 are all different. #

Thus condition (13.9) is sufficient for 0-symmetry in the cases admitted. However, we shall see by many examples that it is not necessary.

The following, rather obvious, consequence of Theorem 21.1 turns out to be a useful tool later on.

COROLLARY 21.1.2:   The Cayley graph of a group satisfying the hypotheses of Theorem 21.1 is 0-symmetric if none of the five nontrivial permutations of the generators induces a graph automorphism.   #

Before continuing with the discussion of these graphs, it will be convenient to distinguish several cases, according to the kinds of groups with three involutory generators that can result from (21.1), (21.2) and (21.3).

(i)   The group J generated by S and Y is cyclic with generator S, where

$$S^n = E . \tag{21.11}$$

Then there is some positive integer k such that

$$Y = S^k . \tag{21.12}$$

For the generators R_i of H, this means that

$$R_1 R_3 = (R_1 R_2)^k , \tag{21.13}$$

or

$$R_3 = (R_2 R_1)^{k-1} R_2 . \tag{21.14}$$

This is the case of a dihedral group D_n of order 2n with one redundant generator.   Examples were already mentioned in Section 13.   A similar case would occur if Y or $S^{-1}Y$ were a generator of J.   In Section 22 we study this case in detail.

(ii)   The group J is cyclic, but neither S nor Y nor $S^{-1}Y$ is a generator of J.   Here, H is again a dihedral group, but now none of the three generators is redundant.   This exceptional case will be studied in Section 23.

(iii)   The group J generated by S and Y is abelian, according to (21.2), but not cyclic.   As has been pointed out by Frucht (1955, p. 9), the group H is then "generalized dihedral".   It can happen that the group H reduces to a direct product of the form $D_n \times C_2$.   (Only even values of n will be of interest, since $D_n \times C_2 \cong D_{2n}$ for n odd, as is well known).   This possibility is not surprising if we apply Theorem 13.2 to the relation (21.14) written in the form

$$(R_2 R_1)^{k-1} R_2 R_3 = E \ . \tag{21.15}$$

This case will be studied in Section 24.

The group $\overline{H}$ generated by three involutory generators satisfying only (21.8) and (21.10) is Coxeter's (1970) group $((p_3, p_1, p_2; 2))$.   (See Section 13.)   The subgroup $\overline{J}$ of $\overline{H}$ that is generated by the products $\overline{S} = R_1 R_2$, $\overline{Y} = R_1 R_3$, satisfying

$$\overline{S}^{p_3} = \overline{Y}^{p_2} = (\overline{S}^{-1}\overline{Y})^{p_1} = E, \ \overline{S} \not\leftrightarrows \overline{Y} \ , \tag{21.16}$$

is the group that Coxeter (1939, pp. 86-88) denotes by $(p_3, p_1, p_2; 1)$.   He shows that the three parameters p_3, p_1, p_2 can be decomposed as

$$p_3 = bcd, \quad p_1 = cad, \quad p_2 = abd \ , \tag{21.17}$$

where d is the greatest common divisor of p_3, p_1, p_2 and

$$(a,b) = (b,c) = (c,a) = 1 \ . \tag{21.18}$$

Furthermore, the group $\bar{J}$ is isomorphic to the direct product

$$\bar{J} \cong C_{abcd} \times C_d \; . \tag{21.19}$$

The numbers a, b, c, d —which we shall call the Coxeter parameters— give new insight into the structure of these groups. Evidently the groups H and J are homomorphic images of $\bar{H}$ and $\bar{J}$. In particular, expression (21.17) implies that J contains an element of order abcd. Hence in cases (i) and (ii) above, where J is cyclic, we must have

$$J \cong C_{abcd} \; ; \tag{21.20}$$

that is, n = abcd.

In case (i), where S is a generator, one of the first three Coxeter parameters (the parameter a, according to (21.17)) must be equal to 1. But in case (ii), on the contrary, all of the first three Coxeter parameters must be greater than 1. On the other hand, in case (iii) the last Coxeter parameter, d, must be greater than 1, and

$$J \cong C_{abcd} \times C_r \tag{21.21}$$

where r > 1 is a factor of d.

It is convenient also to restate Corollary 21.1.1 as follows.

COROLLARY 21.1.3: The Cayley graph of a group satisfying the hypotheses of Theorem 21.1 is 0-symmetric if the three Coxeter parameters a, b, c are all different.   #

We only note that two of p_1, p_2, p_3 are equal if and only if two of a, b, c are equal and thus, by (21.18), equal to 1.

22   CAYLEY GRAPHS OF DIHEDRAL GROUPS WITH
ONE REDUNDANT GENERATOR

In this section we continue the discussion of case (i) from Section 21. Because of the specifications in (21.1), (21.11) and (21.14), we have a dihedral group D_n of order 2n with defining relations

$$R_1^2 = R_2^2 = (R_2R_1)^n = E \qquad (22.1)$$

in which a third involutory—and of course redundant—generator R_3 has been added:

$$R_3 = (R_2R_1)^{k-1}R_2 = R_2(R_1R_2)^{k-1} \qquad (22.2)$$

for some positive integer k < n.

In order to avoid repetitions and (non 0-symmetric) Cayley graphs of girth 4, the following restrictions should be imposed on k:

(i)   2k $\leq$ n+1, because replacing k by n+1-k (equivalent to interchanging generators R_1 and R_2) yields an isomorphic graph;

(ii)   k > 2, since k = 2 implies that $R_2R_1R_2R_3$ = E, and the Cayley graph has girth 4;

(iii)   k $\neq$ [(n+1)/2], since (if n is even) k = n/2 implies $(R_3R_1)^2$ = E, and (if n is odd) k = (n+1)/2 gives $R_3R_2R_3R_1$ = E. In either case the graph has girth 4 again.

Combining these restrictions we see that k should satisfy

$$3 \leq k < n/2 \ . \qquad (22.3)$$

(In the paper by Boreham et al. (1974) —where m is used instead of our k and the graph we are discussing is called G(n,m)—the upper bound of (n+1)/2 in Lemma 2.1 (i), p. 217, is incorrect and should be replaced by n/2.)

Being bipartite, the Cayley graph of our group cannot contain pentagons. Thus it has girth 6, because of the hexagons described by (21.9).

A simple consequence of (22.3) is that n is not less than 7. However, the cases k = 3, n = 7 or 8, correspond to well-known symmetrical (and hence not 0-symmetric) graphs. The first of these is the 6-cage or Heawood graph, with LCF code $[5,-5]^7$. (See Coxeter, 1950, Fig. 9.) The second, with LCF code $[5,-5]^8$, has been described by Coxeter (1950; see Fig. 12) who called it {8}+{8/3}. Since no other value of k satisfies (22.3) for n = 7 or 8, we shall require for the rest of this section that

$$n \geq 9 \ . \qquad\qquad\qquad (22.4)$$

The defining relations (22.1) and (22.2) show that our graph has the Frucht diagram given in Fig. 22.1 (also given by

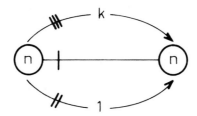

Fig. 22.1   Frucht notation for the dihedral group
         (22.1), (22.2).

Boreham, et al. 1974), in which one, two or three cross-strokes
have been added to the lines corresponding to R_1, R_2 or R_3.
The last relation in (22.1) provides a hamiltonian circuit with
LCF code

$$[2k-1, - (2k-1)]^n . \tag{22.5}$$

We also have the relations

$$R_3 R_1 = (R_2 R_1)^k, \quad R_2 R_3 = (R_1 R_2)^{k-1} \tag{22.6}$$

from which it follows that the parameters of our graph are

$$p_3 = n, \quad p_1 = \frac{n}{(n, k-1)}, \quad p_2 = \frac{n}{(n, k)}, \quad q = 2 . \tag{22.7}$$

Of course, the values of p_1 and p_2 might be interchanged to
satisfy (13.2). The last parameter, q, comes from (21.9).

    The special case, n a prime $p \geq 11$, has been studied by
Boreham et al. (1974), who showed that for $n = p$, the $(p-5)/2$
graphs satisfying (22.3) and (22.5) fall into $[p/6]$ isomorphism classes, and that these graphs are 0-symmetric, with one
exception if $p \equiv 1$ (mod 6). This exceptional case occurs when
k satisfies

$$k(k-1) \equiv -1 \pmod{p} \tag{22.8}$$

and corresponds to a 1-regular graph. By the way, we have
already encountered the two smallest instances, namely

$$p = 13, \ k = 4, \ \text{and} \ p = 19, \ k = 8 , \tag{22.9}$$

in Section 6, where the vertices of these graphs were "blown
up to triangles" in order to obtain graphs of type ^1Z and
girth 3.

Boreham et al. (1974) also studied the case where k = 3
for general n, as mentioned above in Section 13 after (13.12).
Later, Boreham (1974) and Foster (1975) independently took up
the general case, n not necessarily prime. We summarize their
assertions concerning 0-symmetry in a theorem.

THEOREM 22.1:   (Foster)   The graphs of the groups satisfying
(22.1) and (22.2) with k and n satisfying (22.3) and (22.4)
are all 0-symmetric unless one of the following conditions is
fulfilled:

(a) $(k-1)^2 \equiv 1 \pmod{n}$,

(b) $k^2 \equiv 1 \pmod{n}$,

(c) $k(k-1) \equiv -1 \pmod{n}$.

The graph is of type ^3T with t = 1 (see Section 1) in cases
(a) and (b); in case (c), it is of type ^3S and moreover 1-
regular.

We remark that the graphs corresponding to (c) are the
graphs of regular hexagonal maps $\{6,3\}_{b,c}$ on a torus.
(Coxeter and Moser 1980, p. 107.) Indeed, all of the graphs
considered in these sections are embeddable on a torus; that
is, they are derivable from the infinite Euclidean tessellation $\{6,3\}$ by identifying opposite sides of a suitable
parallelogram. As an example, Fig. 22.2 shows the Cayley
graph of the group ((12,6,4;2)) (No. 1 in Table 24.2) so
imbedded.

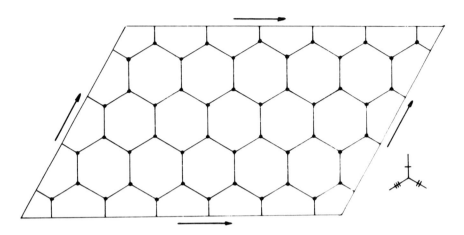

Fig. 22.2   The Cayley graph of the group ((12,6,4;2))
             imbedded on a torus.

PROOF: The proof that conditions (a), (b) or (c) imply non-

0-symmetry is simple.   If p_1 = n (i.e., (k-1,n) = 1) then the

relation $(R_3R_2)^n$ = E provides a second hamiltonian circuit.

There will be a relation of the form

$$(R_3R_2)^{k'} = R_1R_2 \tag{22.10}$$

where k' must satisfy, according to (22.6), the congruence

$$k'(k-1) \equiv -1 \pmod{n} . \tag{22.11}$$

The LCF code for the circuit $(R_3R_2)^n$ = E is

$$[2\underline{k}'-1, -(2\underline{k}'-1)]^n \tag{22.12}$$

where $\underline{k}'$ is the smaller of k' (the least positive solution of

(22.11)) and n+1-k'.   Now, we will have a nontrivial group

automorphism, corresponding to a permutation of the generators

R_1, R_2, R_3 that leaves the graph unchanged, if $\underline{k}'$ = k.   The two possibilities for $\underline{k}'$ in this equation lead to conditions (c) or (a) of the theorem.

Similarly, if $p_2 = n$ (so $(k,n) = 1$) a second hamiltonian circuit is provided by $(R_3R_1)^n = E$, and from a relation of the form

$$(R_3R_1)^{k''} = R_2R_1 \qquad (22.13)$$

we obtain the LCF code

$$[2\underline{k}''-1, \ - \ (2\underline{k}''-1)]^n \qquad (22.14)$$

where $\underline{k}''$ is the smaller of k'' and $n+1-k''$, and k'' is the least positive solution of

$$k''k \equiv 1 \ (\text{mod } n) \ . \qquad (22.15)$$

The two possibilities in the equation $\underline{k}'' = k$ lead to conditions (b) or (c) of the theorem.   #

Proofs that the graphs satisfying none of the conditions (a), (b) or (c) are 0-symmetric will be noted as we determine the isomorphism classes into which they fall, starting with the following theorem.

THEOREM 22.2:   The graph with LCF code $[5,-5]^n$, corresponding to $k = 3$, is 0-symmetric for $n \geq 9$, with isomorphism classes formed by the values of k as shown:

(a) $n \equiv 0$ (mod 6): $k = 3$,

(b) $n \equiv \pm 1$ (mod 6): $k = 3$ or $(n-1)/2$ or $[(n+2)/3]$,

(c) $n \equiv \pm 2$ (mod 6): $k = 3$ or $[(n+2)/3]$,

(d) $n \equiv 3$ (mod 6): $k = 3$ or $(n-1)/2$.

PROOF: As remarked above, the proof of 0-symmetry is given
by Boreham et al. (1974, p. 223) for n $\geq$ 10. For n = 9, the
0-symmetry can be checked individually.

  The isomorphism classes are determined by solving the
congruences (22.11) and (22.15) for k' and k". After setting
k = 3, the congruences become

$$2k' \equiv -1 \quad (\text{mod } n),$$ (22.16)

$$3k'' \equiv 1 \quad (\text{mod } n).$$ (22.17)

The first of these has the solution k' = (n-1)/2 if and only
if n is odd. The second has a solution if and only if
(n,3) = 1. It is easily checked that the solution k" is
either $[(n+2)/3]$ or $n+1-[(n+2)/3]$. In either case, $\underline{k}$" is
$[(n+2)/3]$, as stated in the theorem.   #

  As the next step in studying the isomorphism classes of
the remaining graphs, we determine the number of hexagons
passing through a vertex of the graph. First, we observe
that a walk of length 6 is executed by following the arcs on
the diagram of Fig. 22.1. Let the arc corresponding to R_i be
crossed α_i times in one direction and β_i times in the other.
Which direction is which, is immaterial; so we may assume

$$\alpha_3 \geq \beta_3 .$$ (22.18)

To get length 6 we must have

$$\alpha_1 + \alpha_2 + \alpha_3 = 3, \quad 0 \leq \alpha_i \leq 3;$$ (22.19)

$$\beta_1 + \beta_2 + \beta_3 = 3, \quad 0 \leq \beta_i \leq 3 .$$ (22.20)

The walk returns to the starting point if and only if the congruence

$$\alpha_2 - \beta_2 + k(\alpha_3 - \beta_3) \equiv 0 \quad (\text{mod } n) \tag{22.21}$$

is satisfied. Now $\alpha_3 - \beta_3$ must be one of the integers $0,1,2$ or 3.

(i) $\alpha_3 - \beta_3 = 0$. It follows that $\alpha_2 - \beta_2 = \alpha_1 - \beta_1 = 0$. The possible values for the α's and β's give three hexagons (as mentioned in (21.9)) and several closed walks that are not hexagons and so do not interest us here.

(ii) $\alpha_3 - \beta_3 = 1$. The only possibility consistent with (22.3) is the hexagon $(R_1 R_2)^2 R_3 R_2 = E$, arising when

$$k = 3 . \tag{22.22}$$

(iii) $\alpha_3 - \beta_3 = 2$. The only possibility consistent with (22.3) is the hexagon $(R_3 R_1)^2 R_2 R_1 = E$, arising when n is odd and

$$2k = n - 1 . \tag{22.23}$$

(iv) $\alpha_3 - \beta_3 = 3$. There are several possibilities, summarized below.

$$(R_3 R_1)^3 = E \qquad \text{if } 3k = n , \tag{22.24}$$

$$(R_3 R_1)^2 R_3 R_2 = E \qquad \text{if } 3k = n + 1, \tag{22.25}$$

$$R_3 R_1 (R_3 R_2)^2 = E \qquad \text{if } 3k = n + 2, \tag{22.26}$$

$$(R_3 R_2)^3 = E \qquad \text{if } 3k = n + 3. \tag{22.27}$$

In Theorem 22.2 we have already considered the graphs for which (22.22), (22.23), (22.25) or (22.26) hold. The two remaining cases in which each vertex lies on more than three hexagons are classified in the next theorem.

THEOREM 22.3: Let $n \equiv 0$ (mod 3).
(a) If $n \equiv 0$ (mod 9), the graphs for $k = n/3$ and $k' = (n/3) + 1$ are 0-symmetric and form an isomorphism class.
(b) If $n \equiv 3$ (mod 9) and $k = n/3$, or if $n \equiv 6$ (mod 9) and $k = (n/3) + 1$, the corresponding graph is 0-symmetric and lies in an isomorphism class alone.

PROOF: In the case $n \equiv 3$ (mod 9), $k = (n/3) + 1$, condition (b) of Theorem 22.1 is fulfilled, so the resulting graph is not 0-symmetric. Similarly, for $n \equiv 6$ (mod 9), $k = n/3$, condition (a) of Theorem 22.1 is satisfied.

Now consider the case $3k = n$. According to (21.9) and (22.24), the vertices of the Cayley graph that correspond to generators R_1 and R_3 each lie on three hexagons with E, while that corresponding to R_2 lies on only two and is consequently fixed by the stabilizer of E. If $k > 4$, it can be seen from a drawing of the Cayley graph that the vertex corresponding to $(R_2R_1)^{2k-1}$ is the only one at distance 4 from E that is adjacent to three vertices at distance 3 from E. This vertex is likewise fixed by the stabilizer of E. If $(2k-1,n) = 1$ —i.e., $n \equiv 0$ or 3 (mod 9) —the two elements R_2 and $(R_2R_1)^{2k-1}$ generate the group. Then, by Theorem 4.2, the graph is 0-symmetric. The proof of 0-symmetry is similar for $3k = n + 3$, $k > 5$, and $n \equiv 6$ (mod 9). The case $n = 12$ and $k = 4$ is easily treated by separate argument.

The isomorphism of the two graphs mentioned in (a) above follows from the fact that k and k' satisfy congruence (22.11). On the other hand, for all of the graphs of this theorem, one of the parameters p_1 or p_2 is 3, according to (22.24) and (22.27); thus if n > 9 the graphs of this theorem are not isomorphic to the graph with k = 3, whose parameters can be found from (22.7).   #

We now pass to the cases not covered by Theorems 22.2 and 22.3, i.e., those cases in which the graph contains no hexagons other than those corresponding to relations of the form (21.9).

THEOREM 22.4:   If 3 < k < (n-1)/2 and $3k \notin \{n,n+1,n+2,n+3\}$, the graph with LCF code $[2k-1, -(2k-1)]^n$ is 0-symmetric unless one of the conditions of Theorem 22.1 holds. The isomorphism classes of the 0-symmetric graphs are as follows.

(i)   If (k,n) > 1 and (k-1,n) > 1, k is in an isomorphism class alone.

(ii)   If (k,n) > 1 and (k-1,n) = 1, then k and k' (defined after (22.12)) form an isomorphism class.

(iii)   If (k,n) = 1 and (k-1,n) > 1, then k and k" (defined after (22.14)) form an isomorphism class.

(iv)   If (k,n) = 1 and (k-1,n) = 1, then k, k' and k" form an isomorphism class.

PROOF: For the values of k admitted by the hypotheses, each vertex of the Cayley graph lies on exactly three hexagons. Thus Theorem 21.1 applies: the only nontrivial graph auto-morphism leaving fixed the vertex corresponding to E is a

nontrivial permutation of the generators that carries each
Petrie polygon into one of the same length—i.e., that leads
to an interchange of graph parameters with equal values.

The proof of Theorem 22.1 shows that such a permutation
is a graph automorphism if and only if k satisfies one of
the conditions (a), (b) or (c) of Theorem 22.1. Hence the
graphs fulfilling none of these conditions and satisfying
the hypotheses are 0-symmetric, and only for these have we to
show that the isomorphism classes are as claimed.

(i)   The three parameters are different and neither of the
congruences (21.11), (21.15) has a solution. Thus our graph
lies in an isomorphism class by itself. (Note that 0-symmetry
follows from Corollary (21.1.1).

(ii) We have here $p_1 = p_3 \neq p_2$, so only the interchange of
R_1 and R_3 is possible. Either condition (a) of Theorem 22.1
is satisfied, or else our graph is isomorphic to that resulting from replacing k by $\underline{k}$', found from the solution of (22.11).

(iii) Now we have $p_1 = p_2 \neq p_3$, and the only permissible
permutation of generators is the interchange of R_1 with R_2.
This case is analogous to case (ii), with the difference that
we must use $\underline{k}$", found from the solution of (22.15), in place
of $\underline{k}$'.

(iv) All three parameters are equal. The two interchanges
of generators mentioned above can be used to produce all
possible permutations of generators. Thus the graph, when 0-
symmetric, is in the same isomorphism class with the two
graphs corresponding to $\underline{k}$' and $\underline{k}$".

The proof of 0-symmetry in these cases is analogous to
that given by Boreham et al. (1974).   #

The four theorems of this section allow us to find which of the graphs with LCF code (22.5) are 0-symmetric and their isomorphism classes. It turns out that within the range of our study, $2n \leq 120$, there are 350 non-isomorphic 0-symmetric graphs. They are given, with their isomorphism classes, in Table 22.1 (first calculated by Foster (1975) by hand, and later checked by computer by V. Moll of Santa Maria University).

Explanation of Table 22.1

The columns of Table 22.1, on the three following pages, have these meanings. The first column (NO.) contains an index number. The second column (N) is half the order of the group. In the next three columns, headed K1, K2, K3, are values of k —see (22.2)— that yield isomorphic graphs; recall that k also appears in the LCF code (22.5). The next columns give the graph parameters p_1 and p_2. Since $p_3 = n$, this parameter is not listed. The last three columns contain three of the four Coxeter parameters introduced at the end of Section 21; the remaining parameter is a, equal to 1 for all these groups.

Data table (rotated). Columns: NO. | N | K1 | K2 | K3 | P1 | P2 | C | B | D

Block 1 (NO. 79–117)

NO.	N	K1	K2	K3	P1	P2	C	B	D
79	32	7	10		32	16	28	1	16
80	32	8	9	14	32	4	8	1	4
81	33	3	16		33	1	3	1	1
82	33	4	9		33	13	3	1	13
83	33	5	13		33	1	1	1	1
84	33	6	15		33	1	3	3	1
85	33	7	12	17	33	17	1	1	17
86	33	2	11	13	34	1	2	1	17
87	34	1	7		34	17	2	1	17
88	34	5	8		34	17	2	1	17
89	34	6	15		34	17	2	1	17
90	34	3	14		34	17	2	1	17
91	34	10	12		34	35	1	1	35
92	34	3	9		34	3	5	5	35
93	35	4	10		35	5	7	2	5
94	35	5	14		35	7	5	3	7
95	35	8	6		35	1	3	1	1
96	35	1			35	12	4	4	6
97	35	5	9		35	9	6	2	3
98	35	4			35	6	9	1	9
99	36	9			36	4	2	2	1
100	36	6	8	18	36	4	9	1	28
101	36	9	7	12	36	8	2	5	36
102	36	1		15	36	18	12	2	3
103	36	2	4	16	36	12	3	3	37
104	36	11	3	17	36	9	4	1	37
105	37	1			37	37	1	1	37
106	37	3	13	18	37	37	1	2	37
107	37	4	10	15	37	37	5	3	37
108	37	5	9	16	37	37	7	1	37
109	37	6	7	17	38	1	1	1	19
110	38	8	14		38	19	2	1	19

Block 2 (NO. 40–78)

NO.	N	K1	K2	K3	P1	P2	C	B	D
40	25	4	7	8	25	25	1	1	25
41	25	5	6		25	5	5	1	5
42	25	10	11		25	5	5	1	5
43	26	3	9		26	13	2	1	13
44	26	4	10		26	13	2	1	13
45	26	5	6		26	13	2	1	13
46	26	7	12		26	13	2	1	13
47	26	8	11	11	26	9	3	1	9
48	27	3	17		27	27	3	2	27
49	27	4	8		27	9	1	4	9
50	27	5	10		28	3	3	1	1
51	27	6	9		28	1	9	1	1
52	28	9	9		28	4	2	2	2
53	28	3	2		28	4	4	2	29
54	28	4	11	14	14	29	2	5	29
55	28	5	11	1	7	29	7	1	29
56	29	6		13	29	29	1	2	29
57	29	7	8		29	29	3	3	5
58	29	8	10		29	0	5	1	5
59	29	3	8		30	6	6	1	5
60	29	4	6		30	1	2	1	1
61	30	5	12		30	10	3	2	15
62	30	6		15	15	5	5	1	1
63	30	7	13	10	30	10	6	1	5
64	30	8	1	9	31	3	2	2	5
65	30	3	8	4	31	31	3	3	31
66	30	4	7		31	31	5	1	31
67	31	5	13		31	31	6	1	31
68	31	3	1		32	32	2	1	16
69	31	4	12		32	8	4	1	8
70	32	5	14		32	8	4	1	8
71	32	6			32	16	2	1	16

Block 3 (NO. 1–39)

NO.	N	K1	K2	K3	P1	P2	C	B	D
1	9	3	4		9	3	3	1	3
2	10	3	4	5	10	5	2	1	5
3	11	3	4		11	1	1	1	1
4	12	4			6	4	3	2	2
5	13	3	5	6	13	3	1	3	1
6	14	3	5		14	1	2	1	3
7	15	4	6		15	7	3	1	7
8	16	3	7		16	3	5	3	5
9	17	4	5	8	17	8	5	1	8
10	18	4	5	7	18	4	2	1	4
11	19	4	5		19	7	4	1	1
12	20	4	6		20	6	3	1	1
13	20	5		9	20	9	6	3	3
14	21	5	6	6	20	9	6	1	9
15	21	6	8		20	10	4	1	1
16	22	3	7		21	5	5	1	5
17	22	4	7		21	4	3	1	1
18	23	4	1		22	7	3	2	7
19	23	3	5		22	1	2	1	1
20	24	4	7	10	22	1	2	1	1
21	24	3	8	9	22	1	1	1	1
22	22	2	6	10	22	23	2	2	23
23	23	4	7	9	23	23	3	3	23
24	24	4	9	12	23	8	4	2	4
25	24	10		9	24	6	8	1	2
26	25	3		12	25	3	3	1	1
27	25			25	25	25		1	25

138

NO.	N	K1	K2	K3	P1	P2	C	B	D
118	38	6	5		38	19	2	1	19
119	38	7	1		38	19	2	1	19
120	38	8	1		38	19	2	1	19
121	38	9	12		38	19	2	1	19
122	38	10	17		38	13	3	1	13
123	39	3	8		39	39	3	1	39
124	39	4	10	11	39	13	1	1	13
125	39	5	8		39	39	3	1	13
126	39	6	9		39	13	3	1	13
127	39	7	12		1	13	3	3	1
128	39	16			39	13	4	1	13
129	39	16			39	20	5	1	20
130	40	6	8	20	40	10	5	1	2
131	40	7	14	15	40	8	2	4	4
132	40	5	13	8	40	8	8	2	20
133	40	4		18	40	20	5	1	5
134	40	5	18	19	40	5	8	2	4
135	40	6	17		40	8	5	5	20
136	40	7			20	5	8	1	5
137	41	12	14		41	41	1	1	4
138	41	13	9		41	41	1	1	41
139	41	4	16		41	41	1	2	41
140	41	5	17		41	41	3	2	41
141	42	6	7	20	42	14	2	1	41
142	42	7	16	15	42	21	6	6	7
143	42	8	17	18	7	7	7	2	21
144	42	9			21	6	7	2	7
145	42	10	18	18	21	6	3	2	1
146	42	11	14	19	42	14	3	1	3
147	42	12	9		42	14	2	6	7
148	42	15	16	17	42	7	6	2	21
149	42	16	7	18	42	14	7	2	7
150	42	3	17		42	21	7	1	1
151	42	10	20		21	27	3	3	3
152	42	11	19		21	7	2	2	7
153	42	12			42	14	6	1	7
154	42	15	15	20	42	3	4	2	21
155	42	16	3	21	21	14	3	1	1
156	43	3	21		43	43	1	1	43

NO.	N	K1	K2	K3	P1	P2	C	B	D
157	43	4	1	14	43	43	1	1	43
158	43	5	12	18	43	43	1	1	43
159	43	6	8	17	43	43	1	1	43
160	43	9	16	20	43	43	1	1	43
161	43	10	13	19	44	22	2	1	22
162	44	3	16		44	11	4	1	1
163	44	4	9		44	1	4	1	1
164	44	5	0		44	1	2	1	22
165	44	6	9		44	22	2	1	22
166	44	7	19		22	22	4	2	1
167	44	8	20		21	1	1	4	2
168	44	11			44	4	1	1	1
169	44	12	17		44	4	4	1	1
170	44	13	18		45	25	2	1	22
171	45	14	22		45	15	3	1	15
172	45	8	11		45	9	5	3	9
173	45	4	13		45	9	5	1	3
174	45	5	14	17	45	15	3	1	15
175	45	6	18		45	5	1	5	5
176	45	7	16		45	15	9	3	1
177	45	8			45	5	1	1	3
178	45	9			45	5	5	1	3
179	45	0	16		45	5	5	1	23
180	45	15	15	23	46	23	2	1	23
181	46	2	0	17	46	23	2	1	23
182	46	3	9	19	46	23	2	1	23
183	46	4	14	20	46	23	2	1	23
184	46	5	3		46	23	2	1	23
185	46	6	1		46	23	2	1	23
186	46	7	21		46	23	2	1	23
187	46	8	22		46	23	2	1	23
188	46	11	19		46	23	2	1	47
189	46	12	20		47	47	2	1	47
190	46	18	16		46	47	1	1	47
191	47	3	13	23	46	23	1	1	23
192	47	4	12	17	47	47	1	1	47
193	47	5	13	19	47	47	1	1	47
194	47	6	11	20	47	47	1	1	47
195	47				47	47	1	1	47

NO.	N	K1	K2	K3	P1	P2	C	B	D
196	47	7	9	21	47	47	1	1	47
197	47	10	15	22	47	47	1	1	47
198	47	11	14	18	24	16	3	1	8
199	47	13			16	12	4	2	4
200	48	4	20		48	12	4	3	12
201	48	5	19		16	8	6	1	8
202	48	6			48	6	8	1	2
203	48	9	14		24	16	3	3	8
204	48	10	13		16	24	1	2	24
205	48	12			48	16	12	1	4
206	48	15			24	6	1	2	8
207	48	16			16	3	3	3	1
208	48	21	17	24	24	16	1	2	8
209	48	22	13	16	16	49	1	1	49
210	49		10	12	49	49	1	1	49
211	49		9	11	49	49	1	1	49
212	49	4	8		49	49	7	1	49
213	49	5			49	7	7	1	7
214	49	6	15		49	7	1	1	7
215	49	7	20	23	49	49	2	2	49
216	49	14	22		49	25	5	1	25
217	49	18	17		50	25	5	1	25
218	49	19	18		50	10	5	1	5
219	50	21			50	25	2	2	5
220	50	3			50	25	2	1	5
221	50	4	8		50	5	10	1	25
222	50	5	12		50	25	2	1	5
223	50	6	1		50	5	5	1	5
224	50	7	24		50	25	5	2	25
225	50	9	23		25	50	2	2	5
226	50	10			50	10	2	1	5
227	50	13			50	25	5	1	25
228	50	14	22		25	25	5	1	5
229	50	16	21		50	5	5	2	5
230	50	19	25		50	50	2	1	25
231	50	20	13		51	0	1	1	5
232	51	3	13		51	17	3	1	17
233	51	4	11		51	17	3	1	17
234	51	5		14	51	51	1	1	51

Table (NO. 313–350):

NO.	N	K1	K2	K3	P1	P2	C	B	D
313	58	5	24		58	29	2	1	29
314	58	6	25		58	29	2	1	29
315	58	7	26		58	29	2	1	29
316	58	8	13		58	29	2	1	29
317	58	9	22		58	29	2	1	29
318	58	10	21	29	58	29	2	1	29
319	58	11	27	16	58	29	2	1	29
320	58	12	18	17	58	29	2	1	29
321	58	16	20	23	58	29	2	1	29
322	59	17	15	22	59	59	1	1	59
323	59	1	12	28	59	59	1	1	59
324	59	3	10	27	59	59	1	1	59
325	59	4	11		59	59	1	1	59
326	59	5	18		59	59	1	1	59
327	59	6	14		59	59	1	1	59
328	59	7	24		59	59	3	2	15
329	59	8	26		59	59	5	3	3
330	60	9			30	20	6	4	5
331	60	19			15	15	4	5	15
332	60	25			60	10	4	1	15
333	60	1			60	15	10	3	5
334	60	6	18		20	5	1	3	2
335	60	7	17		20	30	2	1	30
336	60	8			60	4	15	1	2
337	60	9			30	3	1	4	1
338	60	10			50	20	15	3	1
339	60	13			60	5	20	1	10
340	60	14	24		60	30	1	2	1
341	60	15	23		30	4	2	5	6
342	60	16			20	3	15	3	0
343	60	21			30	20	3	2	5
344	60	22			50	5	5	5	1
345	60	26			30	30	3	2	5
346	60	27			30	20	5	5	6
347	60	28			20	15	4	3	15

Table (NO. 274–312):

NO.	N	K1	K2	K3	P1	P2	C	B	D
274	54	17	20	27	54	27	28	1	27
275	54	18	19	18	54	3	18	1	9
276	54	21			27	8	3	2	
277	54	22	19	9	27	18	1	1	55
278	54	3	14		55	55	1	1	1
279	55	4	15		55	55	5	1	1
280	55	6	10		55	55	5	5	55
281	55	7	8		1	1	2	1	1
282	55	1	23	24	55	55	4	1	55
283	55	12	17		55	55	4	1	1
284	55	13	26		55	55	2	5	28
285	55	16	19		1	1	4	1	14
286	55	20	20		24	24	2	1	28
287	56	3	21		48	48	7	2	4
288	56	4	11		28	28	8	7	7
289	56	5			8	8	8	1	28
290	56	6			7	7	2	4	7
291	56	7	25	23	28	28	7	2	28
292	56	8	26		8	8	3	1	2
293	56	9	24		8	8	1	1	4
294	56	10	23		19	19	3	4	49
295	56	7		26	57	57	3	2	19
296	57	8	28		9	9	1	1	57
297	57	18	15		57	57	3	1	9
298	57	21	14		9	9	3	1	57
299	57	22	24		57	57	1	1	9
300	57	3	18		57	57	3	1	57
301	57	4	17		9	9	3	1	9
302	57	5	27		57	57	3	1	57
303	57	6	22		9	9	3	3	9
304	57	7	25		57	57	3	1	29
305	58	10			58	29	9	3	29
306	58	11	20		58	29	2	1	
307	58	12	19		58	58	2	1	

Table (NO. 235–273):

NO.	N	K1	K2	K3	P1	P2	C	B	D
235	51	6	22	23	51	17	3	1	17
236	51	7	10		51	51	1	1	51
237	51	8	20		51	17	3	1	17
238	51	9	19		51	17	3	3	17
239	51	12	15		51	3	3	3	17
240	51	18	24		52	17	2	1	26
241	52	1	17		52	26	4	1	13
242	52	3	21		52	13	4	1	26
243	52	4	22		52	26	2	1	13
244	52	5	16		52	13	4	1	26
245	52	6	23		52	26	4	4	1
246	52	7	19		52	13	2	2	5
247	52	8	20		52	13	1	1	53
248	52	9		26	53	53	1	1	53
249	52	10	18	1	53	53	1	1	53
250	52	11	14	22	53	53	1	1	53
251	53	12	13	16	53	53	1	1	53
252	53	13	9	20	53	53	1	1	53
253	53	14	10	25	53	53	3	2	1
254	53	4	15	24	53	58	3	1	9
255	53	5	14		27	18	2	1	27
256	53	6	23		54	27	6	2	9
257	54	7			54	9	6	1	9
258	54	8			54	27	9	1	27
259	54	12	12		54	6	6	2	3
260	54	13	24		54	6	9	1	9
261	54	4	23		27	27	6	2	
262	54	5		25	54	18	2	1	27
263	54	6		26	54	18	3	2	9

140

We now take up case (ii) of Section 21, in which we assume that the two-letter words

$$R_1R_2 = S, \quad R_1R_3 = Y \tag{23.1}$$

generate a cyclic group J of order n, but neither S nor Y nor $R_2R_3 = S^{-1}Y$ has order n. If we call a generator of that cyclic group T, we may write

$$R_1R_2 = T^h, \quad R_1R_3 = T^k, \quad R_2R_3 = T^{k-h} . \tag{23.2}$$

Fig. 23.1 is a Frucht diagram for the group H in which cross-strokes have been added to the lines to identify the corresponding generators.

Boreham (1974) mentions the study of these graphs as an open problem. Watkins (1974b) discovered the first example of this type: No. 1 in Table 23.1 below. Powers (1982) studies several infinite families of these graphs including all cases with n < 120. This section is a summary of relevant parts of the last reference.

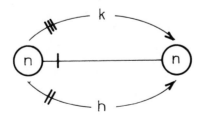

Fig. 23.1 Frucht notation for the dihedral group with no redundant generator.

As we noted at the end of Section 21, this case is characterized by J being cyclic and all of the first three Coxeter
parameters being greater than 1. Since each vertex lies only
on the three hexagons given by (21.9), the graphs are all 0-
symmetric, by Corollary 21.1.3. Moreover, because n = abcd,
n has at least three distinct prime factors; hence the only
values of n falling within the range of our study are 30, 42
and 60.

From (21.17) we must clearly have

$$(n,h) = a, \quad (n,k) = b, \quad (n,k-h) = c \ . \qquad (23.3)$$

Now, replacement of T by T^σ (where $(\sigma,n) = 1$) as generator of
the cyclic group J yields an isomorphic graph, but generally
with different values for h and k. Powers (1982) proves that
this is essentially the only way isomorphism classes are
formed. For n = 30 or 42, there is just one isomorphism
class, while for n = 60 there are two.

In Table 23.1 there are given the LCF codes for four
graphs representing these classes. In the first three cases,
the hamiltonian circuit comes from a relation of the form

$$(R_\lambda R_\mu R_\lambda R_\nu)^{n/2} = E, \quad \{\lambda,\mu,\nu\} = \{1,2,3\} \qquad (23.4)$$

while in the last case the relation is

$$(R_3 R_2 (R_1 R_2)^2)^{20} = E \ . \qquad (23.5)$$

Table 23.1

No.	n	Graph Parameters	Coxeter Parameters	LCF Code
1	30	15,10,6,2	2,3,5,1	$[15,17,-17,-15]^{15}$
2	42	21,14,6,2	2,3,7,1	$[31,33,-33,-31]^{21}$
3	60	20,15,12,2	3,4,5,1	$[15,17,-17,-15]^{30}$
4	60	30,20,12,2	2,3,5,2	$[17,19,-19,19,-19,-17]^{20}$

We now take up case (iii) of Section 21: the group J generated by R_1R_2 and R_1R_3 is abelian, but not cyclic. As we noted at the end of Section 21, J is isomorphic to the direct product $C_{abcd} \times C_r$, where $r > 1$ is a divisor of the fourth Coxeter parameter d. We shall define m as

$$m = abcd/r ,   (24.1)$$

so that the order of the group J is

$$n = mr^2 .   (24.2)$$

The corresponding generalized dihedral group H of order $2mr^2$ can then be generated by three elements R, T, U satisfying the following relations

$$R^2 = T^{mr} = U^r = (RT)^2 = (RU)^2 = T^{-1}U^{-1}TU = E .   (24.3)$$

The case $r = 2$ deserves special mention, for then U commutes with both T and R, and the group H is isomorphic to $D_{2m} \times C_2$, as mentioned already in Section 21. Mark Watkins (1974b) was the first to discover an infinite family of 0-symmetric graphs of this type, with LCF code $[11,13,-13,-11]^{2m}$, whose first members are Nos. 1,2,3,5,6,8,9 and 12 of Table 24.2 below, corresponding to m = 6,9,10,11,12,13,14 and 15. Here as in Section 22, for m = 7 or 8 we find graphs that are ^3T or even ^3S.

Returning to the general case ($r \geq 2$) and in keeping with (21.5) we choose from H the three involutory elements

$$R_1 = R,   R_2 = RT^h,   R_3 = RT^kU .   (24.4)$$

These three do indeed generate the group H if and only if

$$(h,rk,rm) = 1 .$$ (24.5)

From these expressions for the three products

$$R_1R_2 = T^h, \quad R_2R_3 = T^{k-h}U, \quad R_1R_3 = T^kU ,$$ (24.6)

it is easy to calculate the first three parameters of the graph:

$$P_3 = \frac{mr}{(m,h)} , \quad P_1 = \frac{mr}{(m,|k-h|)} , \quad P_2 = \frac{mr}{(m,k)} .$$ (24.7)

Of course we still have q = 2, because of (21.9).

In Section 21 we noted that the graphs of girth 4 are not 0-symmetric. We characterize these graphs in the following.

LEMMA 24.1:   The Cayley graph of the group H generated by the elements (24.4) subject to (24.5) has girth 4 if and only if r = 2 and one of these congruences is satisfied:

   (a) $h \equiv 0$ (mod m) ,

   (b) $k \equiv 0$ (mod m) ,

   (c) $k-h \equiv 0$ (mod m) .

Otherwise the girth is 6.

PROOF: The parameters given in (24.7) are all multiples of r.   Thus one of them can be 2 if and only if r = 2 and one of conditions (a), (b) or (c) is met.   Two of the other four-letter words in the generators have U as a factor.   The other four-letter word is $R_1R_3R_2R_3 = T^{2k-h}U^2$.   For this to equal E would require that r = 2 and $2k-h \equiv 0$ (mod 2m).   But then h would be even, and (24.5) would be violated.

Since the graph is bipartite and contains the hexagons (21.9), its girth must be either 4 or 6. #

In preparation for the application of Theorem 21.1 and its corollaries we characterize those graphs having "extra" hexagons.

LEMMA 24.2: The Cayley graph of the group H generated by (24.4) subject to (24.5) has just the hexagons given by (21.9) unless r = 2 or 3 and one of these is satisfied:

  (a) $h \equiv 0$ (mod m),

  (b) $k \equiv 0$ (mod m),

  (c) $k-h \equiv 0$ (mod m) .

PROOF: If r = 3, each condition is equivalent to one of the parameters p_i in (24.7) being equal to 3. All other six-letter words in the generators are prevented from being equal to the identity by the presence of U or U^2 or by condition (24.5).

If r = 2, one of the six-letter words (different from (21.9)) in which each generator appears twice equals E, precisely when one of the congruences (a), (b) or (c) is satisfied. The remaining six-letter words are prevented from being equal to E by the presence of U or by condition (24.5). #

We now seek a means for obtaining an LCF code for our graphs. To this end we consider the 2r-letter words of the form

$$W_{\lambda\mu\nu} = R_\lambda R_\nu (R_\mu R_\nu)^{r-1}, \quad \{\lambda,\mu,\nu\} = \{1,2,3\} . \tag{24.8}$$

If such a word has order mr and provides a hamiltonian circuit in the graph, an LCF code of the form

$$[2r\alpha-1, \ [2r\alpha+1, \ -(2r\alpha+1)]^{r-1}, \ -(2r\alpha-1)]^{mr} \tag{24.9}$$

is obtained. (The inner bracket means that $2r\alpha+1$ and its negative alternate, each appearing $r-1$ times.) The relation that defines α is

$$R_\lambda R_\mu = W_{\lambda\mu\nu}^\alpha . \tag{24.10}$$

(See the Frucht diagrams for $r = 2$ and 3 in Fig. 24.1.) In order to satisfy Rule 1 of Section 3, we should choose α so that

$$|\alpha| < mr/2 . \tag{24.11}$$

It is convenient, however, to allow the solution of the congruences resulting from (24.10) to range in

$$0 < \alpha < mr \tag{24.12}$$

and agree to replace α by $\alpha-mr$ in (24.9) when (24.11) is violated.

Equation (24.10) can be restated as either of these:

$$E = (R_\lambda R_\mu)^{\alpha-1} (R_\mu R_\nu)^{r\alpha} , \tag{24.13a}$$

$$E = (R_\lambda R_\nu)^{\alpha-1} (R_\mu R_\nu)^{r\alpha-(\alpha-1)} . \tag{24.13b}$$

For any choice of λ, μ and ν, either $R_\lambda R_\mu$ or $R_\lambda R_\nu$ contains U as a factor. Hence we conclude that

$$\alpha \equiv 1 \pmod{r} . \tag{24.14}$$

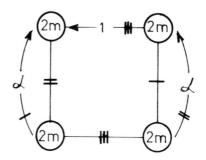

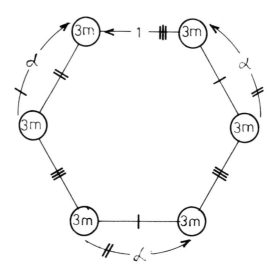

Fig. 24.1   Frucht notation for Cayley graphs of generalized dihedral groups of order $2mr^2$, r=2 and 3.

This condition guarantees that the exponents of U on the
two sides of (24.10) are congruent (mod r).   Furthermore,
the exponents of T on the two sides of (24.10) must be con-
gruent (mod mr).   If the exponent of T in $W_{\lambda\mu\nu}$ is called κ

and that in $R_\lambda R_\mu$ is called ρ, then α must satisfy the congruence

$$\kappa\alpha \equiv \rho \quad (\mathrm{mod}\ mr)\ . \tag{24.15}$$

It is not difficult to verify that— within the range of our study, $mr^2 \le 60$ —any choice of h and k consistent with (24.5) makes at least one of the six words $W_{\lambda\mu\nu}$ provide a hamiltonian circuit in the graph, save in the case

$$m \equiv r-2 \equiv k+h \equiv 0 \quad (\mathrm{mod}\ 3)\ . \tag{24.16}$$

However, the smallest 0-symmetric graph in this family has 144 vertices.

Thus we shall assume that at least one of the words $W_{\lambda\mu\nu}$ provides a hamiltonian circuit in our graph. Moreover we shall choose, for the exponent h in (24.4), the value

$$h = rk+1, \quad 0 \le k \le m-1\ . \tag{24.17}$$

Note that (24.5) is automatically satisfied.

In Table 24.1 are listed the expressions for κ and ρ (for congruence (24.15)) that correspond to this choice.

Table 24.1

$\lambda\ \mu\ \nu$	κ	ρ
1 2 3	$(-r^2+2r)k-r+1$	$rk+1$
1 3 2	$(r^2-r+1)k+r$	k
2 1 3	-1	$-(rk+1)$
2 3 1	$-(2r-1)k-1$	$-(r-1)k-1$
3 1 2	$(r^2-1)k+r$	$-k$
3 2 1	$(-r^2+r-1)k-r+1$	$(r-1)k+1$

It is evident that, under (24.17), W_{213} always provides a hamiltonian circuit, allowing the graph to be described by the LCF code (24.9) with

$$\alpha = h = rk+1 .$$ (24.18)

Thus, all the graphs that can arise (aside from those of (24.16)) are produced when we choose h by (24.17).

THEOREM 24.1: Let the Cayley graph of the group H (24.3) with generators

$$R_1 = R, \quad R_2 = RT^{rk+1}, \quad R_3 = RT^k U$$

have girth 6, and let each vertex lie on just the three hexagons (21.9). Then the graph is 0-symmetric unless one of these congruences holds:

(a) $((r-2)k+1)(rk+1) \equiv 0$ (mod m),

(b) $((2r-1)k+2)k \equiv 0$ (mod m),

(c) $((r+1)k+1)((r-1)k+1) \equiv 0$ (mod m),

(d) $(r^2-r+1)k^2+(2r-1)k+1 \equiv 0$ (mod m) .

REMARKS: (i) Girth 4 occurs if and only if r = 2 and k = 0, m-1, or (for m odd) k = (m-1)/2. Hexagons other than those of (21.9) occur if and only if: the girth is 4, or r = 3 and

$$k \, \varepsilon \, \{0, \ (m-1)/2, \ (m-1)/3, \ (2m-1)/3\}$$ (24.19)

These graphs all fail to be 0-symmetric.

(ii)   The theorem remains correct in the case r = 1. However, because of the different choice of generators, this theorem does not reduce to Theorem 22.1 when r = 1.

PROOF: Under the hypotheses of this theorem, Theorem 21.1 and Corollary 21.1.2 apply. The graph fails to be 0-symmetric if and only if a nonidentical permutation of the generators produces a graph automorphism. Since such a permutation takes word W_{213} into one of the others, a graph automorphism results if and only if $\alpha = rk+1$ is a solution of a congruence (24.15) corresponding to one of the other five words, and that word is of order mr.

Substitution of $rk+1$ for α in each of the five congruences leads to the four conditions of the theorem. (Words W_{132} and W_{321} both give condition (d) because a cyclic change of generators is involved.) It can be shown that each of the congruences of the theorem also implies that the appropriate word has order mr.   #

Isomorphism classes for the 0-symmetric graphs found from Theorem 24.1 are established by choosing k and solving all of the congruences (24.15) for which the corresponding word has order mr. If the same value of k yields more than one α, the corresponding graphs are isomorphic. We note that, for $r = 2$, α and $-\alpha$ are always in the same isomorphism class, so only the positive value is listed in Table 24.2.

Table 24.2*

No.	Order	m	Graph Parameters	a,b,c,d	α
1	48	6	12,6,4	1,2,3,2	3,5
2	72	9	18,18,6	1,1,3,6	3,5,7
3	80	10	20,20,10	1,1,2,10	3
4	80	10	20,10,4	1,2,5,2	5,9
5	88	11	22,22,22	1,1,1,22	3,5,9
6	96	12	24,12,8	1,2,3,4	3,5
7	96	12	24,8,6	1,3,4,2	7,9
8	104	13	26,26,26	1,1,1,26	3,5,7
9	112	14	28,28,14	1,1,2,14	3,11
10	112	14	28,28,14	1,1,2,14	5
11	112	14	28,14,4	1,2,7,2	7,13
12	120	15	30,30,10	1,1,3,10	3,7
13	120	15	30,10,6	1,3,5,2	5,9,11
14	108	6	18,9,6	1,2,3,3	-8,-5,4

*For No. 14, r = 3; for the rest, r = 2. The values of α
listed may be used in the LCF code (24.9).

25   YET ANOTHER 0-SYMMETRIC GRAPH WITH 96 VERTICES

In analogy with Section 21, we now consider a nonabelian group J of order n, with two generators S and Y. Let us suppose that there is an automorphism of J taking each generator to its inverse. We now construct a group H of order 2n by adjoining the involutory element R defined by

$$RSR = S^{-1}, \quad RYR = Y^{-1} . \tag{25.1}$$

The group H can thus be generated by the three involutory elements

$$R_1 = SR, \quad R_2 = R, \quad R_3 = RY \tag{25.2}$$

and may well have a 0-symmetric Cayley graph. Unfortunately, within the range of our study, we have found only one group, not already covered, with a 0-symmetric Cayley graph. We take as J the binary octahedral group called <4,3,2> by Coxeter (1974, p. 68), with defining relations

$$S^4 = Y^3 = (SY)^2 . \tag{25.3}$$

The group H (of order 96) that results from our procedure, called <4,3,2>$_2$ by Coxeter (1974, p. 92), can be generated by the three quaternion transformations

$$\tag{25.4}$$
$$R_\nu : \quad \underline{x} \rightarrow \underline{i} \; \underline{x} \; \underline{q}_\nu \; ,$$

where the quaternions q_ν are

$$\underline{q}_1 = \underline{j}, \quad \underline{q}_2 = \frac{1}{2}(\underline{i} + \sqrt{2}\underline{j} - \underline{k}), \quad \underline{q}_3 = \underline{i} ; \tag{25.5}$$

or by the three unitary matrices:

153

$$R_1 = \begin{pmatrix} 0 & i \\ -i & 0 \end{pmatrix}, \quad R_2 = \frac{1}{2}\begin{pmatrix} -1 & 1+i\sqrt{2} \\ 1-i\sqrt{2} & 1 \end{pmatrix}, \quad R_3 = \begin{pmatrix} -1 & 0 \\ 0 & 1 \end{pmatrix} ;$$

$$(25.6)$$

or by these three matrices over the field of integers mod 17:

$$R_1 = \begin{pmatrix} 0 & 4 \\ -4 & 0 \end{pmatrix}, \quad R_2 = \begin{pmatrix} 8 & 4 \\ -3 & -8 \end{pmatrix}, \quad R_3 = \begin{pmatrix} -1 & 0 \\ 0 & 1 \end{pmatrix} . \qquad (25.7)$$

A presentation is simply

$$R_\nu^2 = E, \quad (R_1R_2)^4 = (R_2R_3)^3 = (R_1R_3)^2 . \qquad (25.8)$$

From the eigenvalues of the products of the matrices in (25.6) it is easy to verify that the parameters of the graph are

$$p_3 = 8, \quad p_1 = 6, \quad p_2 = 4, \quad q = 12 . \qquad (25.9)$$

From the relation

$$(R_1R_3R_2R_1R_2R_3R_2R_3R_2R_1R_2R_3)^8 = E \qquad (25.10)$$

we obtain a hamiltonian circuit in the graph, with the LCF code

$$[-37,9,43,-7,-45,37;-]^8 . \qquad (25.11)$$

The graph is obviously bipartite and is easily seen to be of girth 8. Its 0-symmetry can be proved by means of Theorem 4.2.

TABLES

On the following pages are tabulated all the 0-symmetric graphs found in this study, except the 350 graphs of type ^3Z and girth 6 that are given in Table 22.1. Tables 1 and 2 contain the graphs of types ^1Z and ^3Z respectively. Within each table, graphs are listed by girth, then by order, then arbitrarily. The first column of each table is an index number.

We have followed the text in naming groups, where possible. The symbol GD(p,r) stands for generalized dihedral group with $m = p/r$. (See Section 24.) The extension of a group J by the process of Section 16 is denoted by XJ. The juxtaposition of groups H and K is represented by H&K if different from the direct product H×K. (See Sections 11 and 13.)

In Table 2A, a ditto mark (") in the Group column indicates that the graph is a companion, in the sense of Section 14, of the graph of the last-named group.

A question mark in the column for LCF code means that none has been discovered.

In the last column is a reference to a table (abbreviated T) or a display line in the text where the LCF code or other information about the graph and/or group will be found. Figures (abbreviated F) are cited for those cases where a drawing has been given.

Table 1. Graphs of Type 1_Z

No.	g	ord	group	LCF code	Ref.
1	3	78	$[6,3]^+_{3,1}$	$[2,21,-2;-]^{13}$	(6.2)
2	3	114	$[6,3]^+_{3,2}$	$[2,45,-2;-]^{19}$	(6.3)
3	4	20	$Z(5,2,2)$	$[6,6;-]^5$	T8.2
4	4	40	$Z(10,2,3)$	$[3,13,19,-3;-]^5$	(8.16)
5	4	52	$Z(13,2,5)$	$[10,10;-]^{13}$	T8.2
6	4	68	$Z(17,2,4)$	$[26,26;-]^{17}$	T8.2
7	4	80	$F^{2,-1,1} \& D_4$	$[3,21,27,-3;-]^{10}$	(11.14)
8	4	100	$Z(25,2,7)$	$[14,14;-]^{25}$	T8.2
9	4	104	$Z(26,2,5)$	$[3,37,43,-3;-]^{13}$	(8.17)
10	4	116	$Z(29,2,12)$	$[34,34;-]^{29}$	T8.2
11	4	120	$F^{2,-1,1} \times S_3$	$[3,53,19,-3;-]^{15}$	(11.6)
12	5	110	$Z(11,5,2)$	$[4,-53,25,-17,-4;-]^{11}$	(10.9)
13	5	110	$Z(11,5,7)$	$[4,-43,15,33,-4;-]^{11}$	(10.10)

156

No.	g	ord	group	LCF code	Ref.
14	6	54	Z(9,3,2)	$[5,-11,11,25,-25,-5]^9$	(9.6)
15	6	84	Z(14,3,3)	$[5,21,-29,-19,39,-5;-]^7$	(9.18)
16	6	108	Z(18,3,5)	$[5,-29,29,43,-43,-5]^{18}$	(9.27)
17	7	56	$F^{4,2,1}/C_2$	$[6,24,\infty/1,16,-24,-16,-6]^7$	(12.20)
18	7	112	$F^{4,2,1}$	$[6,24,\infty/6,-33,-24,33,-6]^{14}$	(12.17)
19	8	54	Z(9,3,2)	$[8,27,13,-7,-24,7,-13,24,-8]^6$	(10.3)
20	8	72	$F^{3,2,-1}$	$[7,27,32,-14,-27,14,\infty/3,-7]^8$	(12.4), F12.1
21	8	108	Z(18,3,5)	$[17,-45,31,-7,-51,25,-13,51,19,-19,45,13,-25,-51,7,-31,51,-17]^6$	(10.7)
22	8	112	$F^{4,2,1}$	$[-43,-25,\infty/1,-33,25,33,43]^{14}$	(12.9), F12.2
23	8	112	$F^{4,2,1}$	$[-30,-54,30,7,-27,-22,34,53,-34,27,-7,54,-53,-31,31,22]^7$	(12.12)
24	8	120	$F^{2,-1,1}\times S_3$	$[53,-33,-17,41,9,-53,53,39,7,17,-39,-53,29,-9,-17,-7,33,-29,-19,39,-41,17,-39,19]^5$	(11.9)
25	9	60	Z(5,6,2)	$[18,9,-17,29,-9,18;-]^5$	(9.11)
26	10	100	Z(5,10,2)	$[30,17,-25,-47,-9,9,-33,45,-17,30;-]^5$	(9.23)
27	10	110	Z(11,5,2)	$[35,17,25,-27,35;-]^{11}$	(9.29)
28	10	120	Z(10,6,3)	$[11,-51,-29,17,-9,-59,-37,9,-17,53,-45,-11;-]^5$	(9.38)

Table 2A. Graphs of Type ^3Z and Girth 4

No.	ord	parameters			group	LCF code	Ref.
1	24	4	3	2	3	$G^{3,3,4}$	$[5,-3,12;-]^4$	(15.19)
2	32	8	4	2	8	$B(8,3)$	$[3,13;-]^8$	T17.1
3	36	6	3	2	6	$D_3 \times D_3$	$[5,-3,-17;-]^6$	(18.6)
4	48	4	3	2	6	$G^{3,4,6}$	$[11,-3,7,5,-9,-11;-]^4$	(15.4) F15.1
5	48	6	3	2	4	"	$[7,-3,3,17;-]^6$	(15.7)
6	48	6	4	2	3	"	$[7,-3,-14,-14;-]^6$	(15.10)
7	48	6	4	2	6	$G^{3,4,6}$	$[7,-3,10,10;-]^6$	(15.13)
8	48	12	4	2	6	$B(12,5)$	$[3,21;-]^{12}$	T17.1
9	48	6	4	2	12	"	$[7,-3,11,9;-]^6$	T17.1
10	60	5	3	2	5	$G^{3,5,5}$	$[11,-3,7,5,26,26;-]^5$	(15.22)
11	60	15	6	2	10	$B(15,4)$	$[3,17;-]^{15}$	T17.1
12	60	10	6	2	15	"	$[30,9,7;-]^{10}$	T17.2
13	60	15	10	2	6	"	$[3,-15;-]^{15}$	T17.1
14	64	16	4	2	16	$B(16,7)$	$[3,29;-]^{16}$	T17.1
15	64	8	4	2	8	$D_8 \& D_4$	$[7,-3,11,9;-]^8$	(18.20)

No.	ord	parameters				group	LCF code	Ref.
16	72	6	6	2	6	$D_3 \times D_3 \times C_2$	$[11,-3,15,13,-13,-15;-]^6$	(18.11)
17	80	20	4	2	10	B(20,9)	$[3,37;-]^{20}$	T17.1
18	80	10	4	2	20	"	$[7,-3,11,9;-]^{10}$	T17.1
19	84	21	6	2	6	B(21,8)	$[3,33;-]^{21}$	T17.1
20	84	14	6	2	21	"	$[42,9,7;-]^{14}$	T17.2
21	84	21	6	2	14	"	$[3,-31;-]^{21}$	T17.1
22	96	24	8	2	12	B(24,5)	$[3,21;-]^{24}$	T17.1
23	96	12	8	2	24	"	$[-41,-3,11,9;-]^{12}$	T17.2
24	96	24	12	2	8	"	$[3,-19;-]^{24}$	T17.1
25	96	24	6	2	8	B(24,7)	$[3,29;-]^{24}$	T17.1
26	96	8	6	2	24	"	$[11,-3,19,17,-9,-11;-]^8$	T17.2
27	96	24	8	2	6	"	$[3,-27;-]^{24}$	T17.1
28	96	24	4	2	24	B(24,11)	$[3,45;-]^{24}$	T17.1
29	96	12	4	2	12	$D_{12} \& D_4$	$[7,-3,11,9;-]^{12}$	(18.21)
30	96	12	3	2	8	$XF_4^{1,-2}$	$[7,-3,3,-31;-]^{12}$	(19.20)
31	96	8	3	2	12	"	$[11,-3,7,5,39,37;-]^8$	(19.22)
32	96	12	8	2	3	"	$[-41,-3,34,34;-]^{12}$	(19.23)

159

Table 2A (continued)

No.	ord	parameters				group	LCF code	Ref.
33	96	8	6	2	12	$XF^{1,1,-3}$	$[11,-3,-41,-43,39,37;-]^8$	(19.28)
34	96	12	6	2	8	"	$[-41,-3,3,17;-]^{12}$	(19.30)
35	96	12	8	2	6	"	$[-41,-3,-14,-14;-]^{12}$	(19.31)
36	96	6	4	2	6	$S_4{\times}C_2{\times}C_2$	$[7,41,39,-13,-15,15,13,-7,-9,-39,-41,3,-15,15,-3,9]^6$	(19.36)
37	100	10	5	2	10	$D_5{\times}D_5$	$[9,-3,-25,-27,-49;-]^{10}$	(18.7)
38	108	6	3	2	6	$G^{3,6,6}$	$[17,15,-5,-7,3,-11,-49,-3,-53;-]^6$	(15.16)
39	108	9	6	2	18	$D_9{\times}D_3$	$[17,-3,-41,-43,27,25,-13,-15,-53;-]^6$	(18.15)
40	108	18	6	2	9	"	$[54,33,31,-7,-9,25,23,-15,-17;-]^6$	(18.17)
41	108	18	9	2	6	"	?	(18.13)
42	112	28	4	2	14	$B(28,13)$	$[3,53;-]^{28}$	T17.1
43	112	14	4	2	28	"	$[7,-3,11,9;-]^{14}$	T17.1
44	120	5	3	2	10	$G^{3,5,10}$	$[23,21,-5,-7,-11,-13,13,11,-17,-19,3,-23,23,-3,19,17,13,$ $11,-11,-13,7,5,-21,-23]^5$	(15.28) F15.2
45	120	10	3	2	5	"	$[11,21,19,55,53,26,26,-26,-53,-55,-11;-]^5$	(15.32)
46	120	10	5	2	3	"	$[11,-3,-18,-18,3,25;-]^{10}$	(15.35)
47	120	10	6	2	10	$G^{3,5,10}$	$[59,21,19,-55,-57,36,36,-16,-16,-43,-45,-59;-]^5$	(15.39)

No.	ord	parameters				group	LCF code	Ref.
48	120	30	10	2	6	B(30,11)	$[3,45;-]^{30}$	T17.1
49	120	10	6	2	30	"	$[11,-3,19,17,-9,-11;-]^{10}$	T17.2
50	120	30	6	2	10	"	$[3,-43;-]^{30}$	T17.1
51	120	15	4	2	15	$X(A_4 \times C_5)$	$[7,-3,10,10;-]^{15}$	(19.39)
52	120	6	4	2	5	S_5	$[60,37,35,-7,-9,-15,-17,3,-23,15,-3,23,21,9,7,56,56,-29,-31,$ $60,60,-7,-9,3,-15,23,-3,31,29,17,15,-56,-56,-21,-23,9,7,$ $-35,-37,60]^3$	(19.43)
53	120	6	5	2	4	S_5	?	(19.40)
54	120	5	4	2	6	S_5	?	(19.40)

Table 2B. Graphs of Type ^3Z and Girth 6

No.	ord	parameters				group	LCF code	Ref.
1	48	12	6	4	2	GD(12,2)	$[11,13;-]^{12}$	T24.2, F22.2
2	60	15	10	6	2	D_{30}	$[15,17;-]^{15}$	T23.1
3	72	18	18	6	2	GD(18,2)	$[11,13;-]^{18}$	T24.2
4	80	20	20	10	2	GD(20,2)	$[11,13;-]^{20}$	T24.2
5	80	20	10	4	2	GD(20,2)	$[19,21;-]^{20}$	T24.2
6	84	21	14	6	2	D_{42}	$[31,33;-]^{21}$	T23.1
7	88	22	22	22	22	GD(22,2)	$[11,13;-]^{22}$	T24.2
8	96	24	12	8	2	GD(24,2)	$[11,13;-]^{24}$	T24.2
9	96	24	8	6	2	GD(24,2)	$[27,29;-]^{24}$	T24.2
10	104	26	26	26	2	GD(26,2)	$[11,13;-]^{26}$	T24.2
11	108	18	9	6	2	GD(18,3)	$[23,25,-25;-]^{18}$	T24.2
12	112	28	28	14	2	GD(28,2)	$[11,13;-]^{28}$	T24.2
13	112	28	28	14	2	GD(28,2)	$[19,21;-]^{28}$	T24.2
14	112	28	14	4	2	GD(28,2)	$[27,29;-]^{28}$	T24.2
15	120	30	30	10	2	GD(30,2)	$[11,13;-]^{30}$	T24.2
16	120	30	10	6	2	GD(30,2)	$[19,21;-]^{30}$	T24.2
17	120	30	20	12	2	D_{60}	$[17,19,-19;-]^{20}$	T23.1

No.	ord	parameters				group	LCF code	Ref.
18	120	20	15	12	2	D_{60}	$[15,17;-]^{30}$	T23.1
19	120	6	4	3	5	S_5	?	(20.8)

Table 2C. Graphs of Type ^3Z and Girth 8

No.	ord	parameters				group	LCF code	Ref.
1	60	15	10	10	6	B(15,4)	$[27,9;-]^{15}$	T20.1
2	80	20	10	10	4	B(20,9)	$[-13,29;-]^{20}$	T20.1
3	84	21	14	14	6	B(21,8)	$[-21,9;-]^{21}$	T20.1
4	84	21	14	14	6	B(21,8)	$[15,-39;-]^{21}$	T20.1
5	96	24	8	4	12	B(24,5)	$[27,45;-]^{24}$	T20.1
6	96	24	12	6	8	B(24,5)	$[29,39,31,17,-39,-29;-]^{8}$	T20.1
7	96	24	8	4	6	B(24,7)	$[-21,45;-]^{24}$	T20.1
8	96	8	6	4	12	<4,3,2>$_2$	$[-37,9,43,-7,-45,37;-]^{8}$	(25.11)
9	112	28	14	14	4	B(28,13)	$[19,-35;-]^{28}$	T20.1
10	120	30	10	10	6	B(30,11)	$[27,-51;-]^{30}$	T20.1
11	120	10	6	5	3	$A_5 \times C_2$	?	(20.6)
12	120	10	6	5	10	$A_5 \times C_2$	?	(20.7)
13	120	6	5	4	6	S_5	?	(20.10)
14	120	6	5	4	4	S_5	?	(20.11)

Table 2D. Graphs of Type ^3Z and Girth 10

No.	ord	parameters			group	LCF code	Ref.
1	96	12	8	6	8	B(24,5)	$[35,13;-]^{24}$	T20.1
2	96	12	12	8	4	B(24,11)	$[-13,37;-]^{24}$	T20.1
3	112	28	14	14	4	B(28,13)	$[-13,45;-]^{28}$	T20.1

164

BIBLIOGRAPHY

BIGGS, N.L.

    1974   Algebraic Graph Theory, Cambridge Tracts in Mathe-
            matics, No. 67. Cambridge University Press, London.

BIGGS, N.L. and M.J. HOARE

    1980   A trivalent graph with 58 vertices and girth 9.
            Discrete Math. 30, pp. 299-301.

BOREHAM, T.G.

    1974   Some uses of the paths of graphs in determining their
            groups. Ph.D. Thesis, The University of New
            Brunswick, Fredericton, New Brunswick.

BOREHAM, T.G., I.Z. BOUWER and R.W. FRUCHT

    1974   A useful family of bicubic graphs. In Graphs and
            Combinatorics, R.A. Bari and F. Harary, eds., pp.
            213-224. Lecture Notes in Mathematics, No. 406,
            Springer-Verlag, Berlin.

CAMPBELL, C.M., H.S.M. COXETER and E.F. ROBERTSON

    1977   Some families of finite groups having two generators
            and two relations. Proc. Roy. Soc. London Ser. A
            357, pp. 423-438.

CAPOBIANCO, M. and J.C. MOLLUZZO

    1978   Examples and Counterexamples in Graph Theory, North-
            Holland, New York.

CAYLEY, A.

    1878a On the theory of groups. Proc. London Math. Soc.
            (1) 9, pp. 126-133.

    1878b The theory of groups: graphical representations.
            Amer. J. Math. 1, pp. 174-176.

COXETER, H.S.M.

  1939   The abstract groups $G^{m,n,p}$. Trans. Amer. Math. Soc.
          45, pp. 73-150.

  1950   Self-dual configurations and regular graphs. Bull.
          Amer. Math. Soc. 56, pp. 413-455.

  1970   Twisted Honeycombs. Regional Conference Series in
          Mathematics, No. 4, American Mathematical Society,
          Providence, R.I.

  1973   Cayley diagrams and regular complex polygons. In A
          Survey of Combinatorial Theory, J. N. Srivastava
          et al., eds., pp. 85-93. North-Holland, Amsterdam.

  1974   Regular Complex Polytopes, Cambridge University
          Press, London-New York.

  1975   Faithful trivalent Cayley diagrams. Preliminary
          report. Notices Amer. Math. Soc. 22, No. 720-05-7,
          p. A-39.

  1977   The Pappus configuration and the self-inscribed octa-
          gon. Proc. K. Neder. Akad. Wetensch. A80, pp. 256-
          300.

  1979   On R.M. Foster's regular maps with large faces.
          Proc. of Symposia in Pure Math., Vol. 34, pp. 117-
          127. American Mathematical Society, Providence, R.I.

COXETER, H.S.M. and R. FRUCHT

  1979   A new trivalent symmetrical graph with 110 vertices.
          Ann. N.Y. Acad. Sci., 319, pp. 141-152.

COXETER, H.S.M. and W.O.J. MOSER

  1980   Generators and Relations for Discrete Groups, fourth
          ed.. Springer-Verlag, Berlin-Heidelberg-New York.

FOSTER, R.M.

    1966   A census of trivalent symmetrical graphs I. Presented
           at the Conference on Graph Theory and Combinatorial
           Analysis, University of Waterloo, Waterloo, Ontario.

    1975   Notes on trivalent vertex-transitive graphs.
           Unpublished.

FRUCHT, R.

    1955   Remarks on finite groups defined by generating rela-
           tions. Canad. J. Math. 7, pp. 8-17 and 413.

    1970   How to describe a graph. Ann. New York Acad. Sci.
           175, pp. 159-167.

    1977   A canonical representation of trivalent hamiltonian
           graphs. J. Graph Theory 1, pp. 45-60.

FRUCHT, R., J.E. GRAVER and M.E. WATKINS

    1971   The groups of the generalized Petersen graphs. Proc.
           Cambridge Philos. Soc. 70, pp. 211-218.

GODSIL, C.

    1979   Graphs with Regular Groups, Ph.D. Thesis, University
           of Melbourne, Melbourne, Australia.

GROSSMAN, I. and W. MAGNUS

    1964   Groups and Their Graphs. Random House, New York.

HARARY, F.

    1969   Graph Theory. Addison-Wesley Publishing Co.,
           Reading, Mass.

IMRICH, W.

    1973   On graphical regular representation of groups.
           Colloquia Math. Janos Bolyai 10, Keszthely, 1973,
           pp. 905-925. North Holland, Amsterdam.

1975   On graphs with regular groups. J. Combin. Theory
        Ser. B 19, pp. 174-180.

NOWITZ, L.A. and M.E. WATKINS

  1972a Graphical regular representations of non-abelian
        groups I. Canad. J. Math. 24, pp. 993-1008.

  1972b Graphical regular representations of non-abelian
        groups II. Canad. J. Math. 24, pp. 1009-1018.

POWERS, D.L.

  1982 Exceptional trivalent Cayley graphs for dihedral
        groups. J. Graph Theory 6 (to appear).

TUTTE, W.T.

  1947 A family of cubical graphs. Proc. Cambridge Philos.
        Soc. 43, pp. 459-474.

WATKINS, M.E.

  1971 On the action of non-abelian groups on graphs. J.
        Combin. Theory Ser. B 1, pp. 95-104.

  1974a Graphical regular representations of alternating,
        symmetric and miscellaneous small groups.
        Aequationes Math. 11, pp. 40-50.

  1974b Private communication to H.S.M. Coxeter.

Index